Prospects of Artificial Intelligence in the Environment

Ajitanshu Vedrtnam · Kevin P. Wyche ·
Balendra V. S. Chauhan · Sneha Verma

Prospects of Artificial Intelligence in the Environment

Springer

Ajitanshu Vedrtnam
Mineral and Energy Economy Institute
Polish Academy of Sciences
Kraków, Poland

Institute of Construction and Architecture
Slovak Academy of Sciences
Bratislava, Slovakia

Department of Mechanical Engineering
Invertis University
Bareilly, UP, India

Balendra V. S. Chauhan
Centre for Earth Observation Science
University of Brighton
Brighton, UK

Kevin P. Wyche
Centre for Earth Observation Science
University of Brighton
Brighton, UK

Sneha Verma
School of Science and Technology
City University of London
London, UK

ISBN 978-981-96-6862-5 ISBN 978-981-96-6863-2 (eBook)
https://doi.org/10.1007/978-981-96-6863-2

With Contributions by Akshay Garg, Abuzar, Siddharth Jain—Chap. 3; Tanmoy Banerjea, M. K. Shukla—Chap. 7

This Springer imprint is published by the registered company Springer Nature Singapore Pte Ltd.
The registered company address is: 152 Beach Road, #21-01/04 Gateway East, Singapore 189721, Singapore

Competing Interests The authors have no competing interests to declare that are relevant to the content of this manuscript.

Contents

1 Introduction to the Prospects of Artificial Intelligence in the Environment ... 1
Ajitanshu Vedrtnam, Kevin P. Wyche, Balendra V. S. Chauhan, and Sneha Verma
1 Introduction ... 2
2 Mathematical Modeling in AI for Environmental Science ... 7
2.1 Introduction to Mathematical Modeling in AI for Environmental Science ... 7
2.2 Core Mathematical Concepts in AI for Environmental Science ... 13
2.3 Machine Learning and Mathematical Formulations ... 17
2.4 Numerical Methods and AI-Driven Simulations ... 20
3 Conclusion ... 22
References ... 23

2 Review on Artificial Intelligence in the Environmental Monitoring ... 29
Balendra V. S. Chauhan, Ajitanshu Vedrtnam, Kevin P. Wyche, and Sneha Verma
1 Introduction ... 30
2 Overview of Environmental Monitoring ... 34
2.1 Traditional Methods of Environmental Monitoring and Their Limitations ... 34
3 Role of Artificial Intelligence in Environmental Monitoring ... 38
3.1 Transformative Impact of AI on Environmental Surveillance ... 38

3.2 Advanced Data Analysis and Prediction Techniques in AI for Environmental Monitoring 40
4 Application of AI in Various Fields of Environmental Monitoring 47
5 Challenges Associated with Implementing AI in Environmental Monitoring 47
6 Conclusion 54
References 54

3 Optimizing Renewable Energy with AI: Environmental Monitoring and Predictive Maintenance Strategies 61
Akshay Garg, Abu Zar, Balendra V. S. Chauhan, and Siddharth Jain
1 Introduction 62
2 Renewable Energy Systems 64
2.1 Solar Energy Systems 64
2.2 Environmental Impacts of Solar Energy 65
2.3 Wind Energy Systems 65
2.4 Hydro Energy Systems 67
2.5 Biomass Energy Systems 69
2.6 SWOT Analysis 70
3 Role of AI 75
3.1 Role of AI in the RE Systems 75
3.2 Application of AI in RE Systems 76
4 Limitation 87
4.1 Handling Large Datasets 87
4.2 Integration with Existing Infrastructure 88
4.3 Cybersecurity 88
4.4 Regulations and Policies 89
5 Conclusion 89
References 90

4 Amalgamation of Artificial Intelligence and Climate Change 97
Balendra V. S. Chauhan, Ajitanshu Vedrtnam, Kevin P. Wyche, and Sneha Verma
1 Introduction 98
2 Overview of Climate Modeling 100
2.1 Introduction to Climate Modelling 100
2.2 Types of Climate Models 102
2.3 Components of Climate Models 105
2.4 Challenges in Climate Modeling 108
3 Predictive Climate Modeling Using AI 111
3.1 Introduction to AI in Climate Science 111
3.2 AI Techniques for Climate Prediction 115
3.3 Hybrid AI Models 116
3.4 Data Requirements and Challenges 118
4 AI-Driven Climate Prediction Models: Case Studies 119

4.1 Climate Downscaling Using Neural Networks 119
4.2 Predicting El Niño Events with Machine Learning 121
4.3 Hybrid AI for Sea-Level Rise Prediction 123
4.4 Extreme Weather Event Prediction 125
5 Ethical and Policy Implications of AI in Climate Prediction 126
6 Model Validation, Uncertainty Quantification, and Future Directions 132
6.1 Model Validation Techniques 132
6.2 Uncertainty Quantification 132
6.3 Future Research Directions 133
7 Conclusion 134
References 134

5 Revolutionizing Natural Resource Management with Artificial Intelligence: A Review 143
Balendra V. S. Chauhan, Ajitanshu Vedrtnam, Kevin P. Wyche, and Sneha Verma
1 Introduction 144
1.1 AI Technologies in Natural Resource Management 145
2 Applications of AI in Natural Resource Management 148
2.1 Water Resource Management 148
2.2 Forest Management 149
2.3 Mineral and Mining Management 152
2.4 Biodiversity Conservation 154
3 Benefits of AI in Natural Resource Management 157
3.1 Improved Efficiency and Accuracy 157
3.2 Enhanced Data Utilization 159
3.3 Cost Savings 159
3.4 Increased Accessibility 160
4 Challenges and Considerations 160
4.1 Data Quality and Availability 160
4.2 Ethical and Privacy Concerns 161
4.3 Integration with Existing Systems 161
4.4 Skill Requirements 162
5 Future Directions 162
5.1 Advancements in AI Techniques 163
5.2 Integration with IoT 163
5.3 Collaboration and Partnerships 164
6 Conclusion 164
References 165

6 AI for Natural Disaster Prediction and Management 171
Balendra V. S. Chauhan, Ajitanshu Vedrtnam, Kevin P. Wyche, and Sneha Verma
1 Introduction 172
2 AI Techniques for Earthquake Prediction and Early Warning 173

2.1 Machine Learning Models for Earthquake Prediction 173
2.2 Pattern Recognition and Anomaly Detection 175
2.3 Integration of AI with Seismic Monitoring Systems 176
2.4 Predictive Models for Aftershocks 177
2.5 Challenges and Future Directions 178
3 AI Techniques for Flood Prediction and Management 179
3.1 Machine Learning Models for Flood Forecasting 179
3.2 Remote Sensing and AI Integration 181
3.3 Real-Time Flood Monitoring and Early Warning Systems 182
3.4 Flood Risk Assessment and Management 183
3.5 Post Flood Damage Assessment and Recovery 184
4 AI Applications in Hurricane and Typhoon Management 186
4.1 Enhanced Hurricane and Typhoon Forecasting with AI 187
4.2 Satellite Data Analysis 188
4.3 Real-Time Storm Tracking and Prediction 189
4.4 Risk Assessment and Impact Prediction 190
4.5 Post-Storm Damage Assessment and Recovery 191
5 AI in Managing Volcanic Eruptions 192
5.1 Predictive Models for Volcanic Eruptions 192
5.2 Remote Sensing and AI Integration 194
5.3 Real-Time Monitoring and Early Warning Systems 195
5.4 Impact Assessment and Response Optimization 196
5.5 Risk Assessment and Management 197
6 Future Directions of AI in Natural Disaster Management 198
6.1 Integration with Emerging Technologies 198
6.2 Collaboration and Knowledge Sharing 199
7 Conclusion 200
References 201

7 The Dark Side of AI: Negative Environmental Impacts of AI Advancements 209
Tanmoy Banerjea, M. K. Shukla, Balendra V. S. Chauhan, and Sneha Verma
1 Introduction 210
2 Electronic Waste 211
3 Resource Consumption 214
3.1 Energy 214
3.2 Water 216
4 Self Driving Cars 217
4.1 Connected and Autonomous Vehicle 217
4.2 Fully Automated Vehicle 218
5 Added Pressure on REEs 221
6 Conclusion 223
References 224

Chapter 1
Introduction to the Prospects of Artificial Intelligence in the Environment

Ajitanshu Vedrtnam, Kevin P. Wyche, Balendra V. S. Chauhan, and Sneha Verma

Abstract Artificial Intelligence (AI) possesses immense potential for addressing a wide spectrum of environmental challenges and facilitating sustainable practices. Within the realm of Environmental Science, AI has emerged as a transformative force, offering innovative solutions to pressing ecological concerns. This comprehensive book delves into the myriad applications of AI in Environmental Science, spotlighting its pivotal role in diverse domains. The initial segment of the book provides a succinct overview of its contents, setting the stage for an in-depth exploration of AI's environmental applications. The subsequent sections spotlight AI-driven modelling, prediction, and monitoring techniques, leveraging environmental data across various sectors. Employing AI-powered sensors and remote sensing technologies, substantial volumes of environmental data, encompassing air and water quality, deforestation rates, and climate dynamics, can be harnessed and analysed. Machine learning algorithms process this information to discern trends, anomalies, and forecast potential environmental issues. The book's third section concentrates on the integration of AI into environmental management. Ecosystem preservation, natural disaster mitigation, waste management, and resource conservation take centre stage.

A. Vedrtnam (✉)
Mineral and Energy Economy Institute, Polish Academy of Sciences, Kraków, Poland
e-mail: ajitanshu.m@invertis.org

K. P. Wyche · B. V. S. Chauhan
Centre for Earth Observation Science, University of Brighton, Brighton, UK
e-mail: k.p.wyche@brighton.ac.uk

B. V. S. Chauhan
e-mail: b.chauhan@brighton.ac.uk

B. V. S. Chauhan · S. Verma
School of Engineering, Newcastle University, Newcastle Upon Tyne, United Kingdom

A. Vedrtnam
Institute of Construction and Architecture, Slovak Academy of Science, Bratislava, Slovakia

Department of Mechanical Engineering, Invertis University, 243001 Bareilly, India

S. Verma
e-mail: Sneha.verma@newcastle.ac.uk

A. Vedrtnam et al., *Prospects of Artificial Intelligence in the Environment*,
https://doi.org/10.1007/978-981-96-6863-2_1

AI's prowess in ecosystem analysis, encompassing biodiversity, species distribution, and habitat alterations, enables informed conservation decisions and sustainable land and resource administration. Harnessing historical data, AI algorithms predict natural calamities ranging from hurricanes to wildfires thus facilitating timely alerts and effective disaster response strategies. Diverse themes form the purview of the fourth section, embracing facets such as environmental impact assessment and carbon footprint reduction. Here, AI's automation capabilities shine through in evaluating the environmental repercussions of projects spanning construction, infrastructure development, and urban planning. By delineating potential consequences and proposing mitigation tactics, AI optimizes environmentally conscious decision-making. Moreover, the book delves into cutting-edge topics like AI-driven reduction of carbon footprints and the application of AI in agricultural data processing for enhanced crop yield. Precision agriculture benefits from AI's analysis of satellite, drone, and sensor data to fine-tune crop yields, manage irrigation, and curtail pesticide employment. Ultimately, this insightful tome caters to a broad spectrum of readers—ranging from academicians, students, and researchers to ardent AI enthusiasts operating within the sphere of environmental science. As AI continues its meteoric rise, its symbiotic relationship with environmental science emerges as an indispensable instrument for a sustainable future.

Keywords Artificial intelligence · Environment · Environmental monitoring · Climate modeling · Prediction · Ecosystem management · Natural disaster management · Renewable energy optimization · Waste management · Conservation · And wildlife protection

1 Introduction

AI is important for a variety of reasons due to its potential to transform and enhance various aspects of our lives and industries. AI has become increasingly important in various aspects of our lives due to its ability to process vast amounts of data, learn from patterns, and perform tasks that typically require human intelligence. It can be used to monitor and manage environmental data, predict natural disasters, and optimize resource consumption, contributing to sustainable practices. In the realm of environmental science, AI stands as a transformative tool with the potential to revolutionize various facets of research and monitoring. Through the amalgamation of AI-driven technologies such as machine learning, data analytics, and predictive modeling, environmental scientists can harness the power of large and complex datasets. AI algorithms can seamlessly process diverse data sources, including satellite imagery, sensor networks, and geospatial information, enabling comprehensive and real-time assessment of environmental conditions [1–4]. These capabilities extend to precise monitoring of air and water quality [5, 6], wildlife behavior [7], climate trends [8], and natural resource management [9]. Furthermore,

AI's adeptness in anomaly detection facilitates the early identification of environmental disturbances, allowing for timely intervention and mitigation. By embracing AI in environmental science, researchers can unlock deeper insights, enhance accuracy in predictions, and equip policymakers with informed strategies to address ecological challenges and foster sustainable practices.

The first chapter introduces all the concepts used in this book. It sets the stage for a detailed exploration of how AI technologies can be harnessed to address a wide range of environmental challenges. The subsequent chapters build on this foundation, delving into specific areas where AI is making significant impacts, from environmental monitoring and modeling to management and conservation efforts. This serves as a crucial primer for understanding the scope and potential of AI in fostering a more sustainable interaction with our natural environment.

Further chapters of the book includes the strategies in which AI can be used in environmental modelling, climate modeling and prediction, air and water quality monitoring. The chapters discuss the fact that environmental monitoring has witnessed a transformative paradigm shift with the integration of AI methodologies. AI has demonstrated its prowess in revolutionizing data collection, analysis, and interpretation in the realm of environmental assessment [10]. Leveraging advanced sensors, satellite imagery, and IoT devices, AI algorithms can efficiently process vast volumes of heterogeneous data, extracting meaningful patterns and insights [11]. Machine learning techniques such as neural networks, decision trees, and clustering algorithms empower predictive modeling, enabling accurate forecasts of air quality [12], water contamination [13], biodiversity shifts [14], and natural disaster occurrences [15]. Furthermore, AI-driven anomaly detection swiftly identifies aberrations from expected environmental norms, facilitating prompt responses to potential crises. The fusion of AI and environmental monitoring not only enhances the precision and scope of data analysis but also empowers policymakers and stakeholders to make informed decisions, steering us towards a more sustainable and ecologically balanced future.

Climate modeling and prediction have experienced a paradigm shift with the integration of AI offering unprecedented opportunities to enhance the accuracy and scope of our understanding. By leveraging AI-driven techniques like machine learning, neural networks, and ensemble modeling, climate scientists can effectively analyze vast and intricate climate datasets [8]. AI algorithms excel at discerning intricate patterns and nonlinear relationships within these datasets, enabling the development of more robust and sophisticated climate models. These models, empowered by AI, can simulate complex interactions between various components of the Earth system, including atmosphere, oceans, and land, thereby yielding more reliable and precise climate projections. Additionally, AI-driven predictive capabilities enable the identification of potential tipping points and extreme events, enhancing our ability to anticipate and adapt to changing climatic conditions. In essence, the fusion of AI and climate modeling holds immense promise in advancing our comprehension of Earth's intricate climate dynamics and in formulating strategies for sustainable climate management and policy formulation [16].

The integration of AI has ushered in a new era of innovation and accuracy in air and water quality monitoring. AI offers a transformative approach to handling the complex and vast datasets involved in environmental assessment. Through advanced sensor networks, satellite imagery, and IoT devices, AI algorithms can efficiently process diverse data sources, identifying subtle patterns and correlations that might elude conventional methods. Machine learning techniques, such as deep learning and regression analysis, enable precise modeling of air and water pollution dynamics, facilitating early detection of contamination sources and the formulation of effective mitigation strategies. AI's real-time processing capabilities provide timely alerts and insights, enhancing our ability to safeguard human health and ecosystems. Moreover, AI-powered predictive modeling enables us to anticipate future air and water quality trends, aiding policy decisions aimed at sustainable environmental management [17–20]. The marriage of AI and air and water quality monitoring holds immense potential for elevating the precision, scope, and responsiveness of environmental assessment and protection.

The book includes chapters which describe the applications of AI in environmental management techniques. The topics included mainly focuses on AI based ecosystem management, natural disaster prediction and management, waste management, and natural resource management. The application of AI has introduced a transformative dimension to ecosystem management, offering novel avenues for comprehensive understanding and sustainable stewardship. By harnessing AI-driven technologies such as machine learning and remote sensing, ecosystem managers can process vast and heterogeneous datasets with unprecedented efficiency and accuracy. AI algorithms adeptly uncover intricate patterns and ecological relationships within these datasets, facilitating species distribution modeling, habitat assessment, and biodiversity monitoring [21, 22]. Moreover, AI-powered predictive analytics enable the projection of ecosystem responses to varying stressors and management scenarios, aiding in the formulation of adaptive strategies. Real-time data integration and anomaly detection empower timely identification of disturbances, enabling swift intervention and restoration efforts. Through AI's dynamic and iterative capabilities, ecosystem management can transcend traditional boundaries, fostering resilient ecosystems and guiding informed decision-making to ensure the long-term health and balance of our natural world.

The integration of AI marks a groundbreaking advancement in natural disaster prediction and management [23–25], revolutionizing our capacity to mitigate and respond to catastrophic events. AI's prowess in processing vast and diverse datasets, including satellite imagery, weather patterns, and seismic data, equips us with unprecedented insights into impending disasters. Through machine learning algorithms and predictive modeling, AI can identify subtle indicators and nonlinear relationships that elude traditional methods, enabling accurate forecasting of events like hurricanes, earthquakes, and floods. Furthermore, AI's real-time analysis and anomaly detection provide timely alerts and early warnings, facilitating swift evacuation and resource allocation. In disaster aftermath, AI aids in damage assessment, utilizing image recognition and spatial analysis to expedite recovery efforts. By augmenting human expertise with AI-driven insights, natural disaster prediction and management enter a new era of efficacy, enhancing our resilience and adaptive strategies in the face of nature's most formidable challenges.

The incorporation of AI has catalyzed a transformative shift in waste management practices [26], redefining how we approach the challenges of waste reduction, recycling, and resource optimization [27]. AI's data-driven capabilities, coupled with machine learning algorithms, have revolutionized waste sorting, and recycling processes, enabling automated identification and separation of diverse materials with remarkable accuracy. Through computer vision and robotics, AI-driven systems enhance the efficiency of waste collection and sorting operations, minimizing contamination, and maximizing resource recovery [28]. Furthermore, AI's predictive analytics empower waste management strategies, offering insights into waste generation patterns and optimizing collection routes for minimal environmental impact. Real-time monitoring and anomaly detection enable prompt identification of irregularities or overflow, facilitating timely interventions. By converging AI with waste management, we embark on a sustainable trajectory, streamlining operations, reducing landfill burdens, and fostering circular economies that preserve valuable resources for generations to come.

The book covers miscellaneous areas of how AI can be used in areas such as renewable energy optimization, conservation and wildlife protection, agriculture, and land use, and carbon footprint reduction. The integration of AI has ushered in a transformative era for renewable energy optimization, offering unparalleled avenues to enhance efficiency, reliability, and sustainability in the generation and utilization of clean energy sources. AI's advanced algorithms and machine learning techniques play a pivotal role in real-time energy forecasting, demand prediction, and grid management, facilitating the integration of variable renewable sources like solar and wind into the energy mix [29, 30]. Through AI-driven analytics, optimal operational strategies for renewable energy systems can be formulated, maximizing power generation while minimizing intermittency-related challenges. Furthermore, AI-enabled predictive maintenance enhances the performance and lifespan of renewable energy infrastructure, ensuring minimal downtime and reduced maintenance costs. Additionally, AI-driven resource assessment aids in identifying ideal locations for renewable energy installations, optimizing energy output. By synergizing AI with renewable energy, we pave the way for a sustainable energy landscape that effectively addresses climate change while meeting the world's growing energy demands [30].

The incorporation of AI marks a revolutionary stride in conservation and wildlife protection, redefining strategies to safeguard biodiversity and ecological balance [31, 32]. AI's data-driven capabilities, bolstered by machine learning algorithms, have transformed the landscape of wildlife monitoring and management [33]. Through advanced image recognition, acoustic analysis, and sensor technologies, AI-driven systems enable efficient species identification and tracking, facilitating the collection of critical data on population dynamics and migration patterns. Moreover, AI's predictive modeling enhances our understanding of habitat suitability, aiding in the formulation of informed conservation plans and restoration efforts. AI's real-time monitoring capabilities revolutionize anti-poaching initiatives, enabling swift response to threats and enabling rangers to focus efforts where they are most needed [34]. Additionally, AI-powered algorithms process vast datasets from satellite imagery and remote sensors, uncovering insights into deforestation, habitat loss, and illegal activities, which pose imminent dangers to wildlife and ecosystems. Additionally, the integration of AI extends beyond monitoring and enforcement. It empowers data-driven decision-making, optimizing resource allocation for conservation projects, and evaluating the effectiveness of various interventions. AI-driven predictive analytics forecast potential conflicts between human activities and wildlife habitats, aiding in proactive measures to minimize negative interactions. Furthermore, AI-driven ecological modeling allows researchers to simulate various scenarios, projecting the potential impact of climate change and other stressors on species and ecosystems. This proactive approach informs adaptive strategies and policy recommendations, fortifying the resilience of wildlife populations in the face of evolving environmental challenges. In essence, the marriage of AI and conservation ushers in a new era of efficacy and precision, amplifying our ability to protect and restore endangered species and delicate ecosystems. As we navigate an increasingly complex and interconnected world, AI stands as an indispensable ally in our relentless pursuit to preserve the rich tapestry of life on Earth.

The integration of AI has ignited a transformative paradigm in agriculture and land use, revolutionizing how we optimize resource allocation, enhance productivity, and ensure sustainable practices [35, 36]. AI's data-driven capabilities, coupled with machine learning algorithms, have redefined precision agriculture, enabling real-time analysis of soil health, crop growth, and pest infestations [37]. Through satellite imagery and remote sensors, AI-driven systems provide comprehensive insights into land conditions, facilitating informed decisions on irrigation, fertilization, and crop rotation for maximized yields while minimizing environmental impact. AI-powered predictive modeling forecasts weather patterns and climate changes, aiding farmers in adapting planting and harvesting schedules to mitigate risks. Furthermore, AI's data analytics enhance supply chain efficiency, optimizing distribution and reducing food waste. The integration of AI extends beyond individual farms, aiding in land use planning and management [38]. AI-driven geospatial analysis assists in identifying suitable locations for cultivation and conservation, ensuring responsible land use and preventing habitat degradation. Moreover, AI's capacity to process vast datasets aids in monitoring deforestation, land degradation, and urban expansion, providing valuable insights for policy decisions and sustainable land management. AI also plays a pivotal role in disease detection and management, identifying crop diseases early through image recognition and predictive algorithms, enabling timely intervention and reducing crop losses [39]. To sum up the chapters included in this section shows that the fusion of AI with agriculture and land use revolutionizes traditional practices, offering a holistic approach that optimizes resource utilization, enhances productivity, and promotes sustainable stewardship of our land and natural resources. As we navigate the challenges of a changing climate and growing global population, AI emerges as a transformative force in securing food security, conserving biodiversity, and ensuring the vitality of our planet's ecosystems.

The incorporation of AI heralds a transformative era in the pursuit of mitigating carbon footprints, offering a diverse array of innovative approaches to drive sustainable practices and minimize environmental impact [40, 41]. AI's data-driven capabilities, complemented by machine learning algorithms, are harnessed to optimize energy consumption, streamline transportation systems, and enhance industrial processes. Through real-time monitoring and analytics, AI identifies energy inefficiencies and recommends tailored solutions for energy conservation and emission reduction. Smart grids, enabled by AI, facilitate dynamic energy distribution and demand response, contributing to a more efficient and eco-friendly energy ecosystem. AI-powered predictive models analyze complex data sets to forecast energy demand patterns, enabling better grid management and allocation of renewable resources.

In the realm of transportation, AI optimizes route planning, vehicle performance, and traffic flow, minimizing fuel consumption and emissions [42]. Furthermore, AI supports the development of sustainable materials and manufacturing processes, enhancing the circular economy and reducing the carbon intensity of products [43]. AI-driven simulations and virtual modeling aid in designing energy-efficient buildings and urban environments, reducing energy needs for heating, cooling, and lighting. The integration of AI in carbon footprints transcends sectors, providing a comprehensive toolkit to inform policy decisions, business strategies, and individual behaviors in pursuit of a more sustainable future. By capitalizing on AI's potential, we can recalibrate human activities towards a low-carbon or high carbon trajectory, forging a path towards climate resilience and ecological equilibrium. The book as a whole covers most of the prospects of artificial intelligence in AI, and we strongly feel that it would act as one of the directories for further researchers in the field of environmental science with implications of artificial intelligence.

2 Mathematical Modeling in AI for Environmental Science

Mathematical models serve as the backbone of AI-driven environmental applications, translating complex ecological and atmospheric phenomena into quantifiable insights.

2.1 Introduction to Mathematical Modeling in AI for Environmental Science

Mathematical modeling has long been a crucial tool in environmental science, providing a systematic approach to understanding complex ecological, atmospheric, and hydrological processes. These models use mathematical equations to describe environmental systems, enabling scientists to predict future trends, analyze interactions, and develop strategies for sustainable management. With the emergence of AI, mathematical modeling has undergone significant advancements, incorporating machine learning, deep learning, and statistical algorithms to improve accuracy, efficiency, and real-time adaptability [44]. This section explores the definition and significance of mathematical modeling in AI-driven environmental research and examines how AI enhances traditional modeling techniques for better prediction and decision-making.

2.1.1 Definition and Importance of Mathematical Modeling in AI-Driven Environmental Research

Mathematical models in environmental science are structured frameworks that describe relationships between different environmental variables, helping researchers quantify processes such as air pollution dispersion, climate change effects, and water resource management. These models are classified as:

Empirical Models: Empirical models are based on statistical relationships derived from observed data rather than fundamental physical laws. These models rely on historical environmental data to identify patterns and correlations between variables, making them particularly useful in air quality forecasting, climate trend analysis, and hydrological studies [45, 46]. For example, in air pollution prediction, empirical models use regression techniques to estimate pollutant concentrations based on meteorological conditions (e.g., temperature, humidity, wind speed) and emission sources. A common empirical model is the multiple linear regression (MLR) equation:

$$C_{pollutant} = \beta_0 + \beta_1 T + \beta_2 RH + \beta_3 WS + \epsilon$$

where, $C_{pollutant}$ is the concentration of a pollutant (e.g., NO_2, $PM_{2}.{}_5$); T is temperature, RH is relative humidity, WS is wind speed, β_0, β_1, β_2, β_3 are the regression coefficients, and ϵ represents the error term. Empirical models are computationally efficient and easy to implement but may struggle with generalizability, especially when applied to new or unseen environmental conditions. AI techniques, such as neural networks and support vector machines, can enhance empirical models by capturing nonlinear relationships and improving predictive accuracy for not only air quality prediction but also for various applications like solar irradiation prediction [47].

Mechanistic Models: Mechanistic models, also known as process-based models, are built on physical, chemical, and biological principles to simulate environmental processes. These models describe cause-and-effect relationships through mathematical equations and are widely used in weather forecasting, air pollution transport modeling, and hydrodynamic simulations. One of the most used mechanistic models in environmental science is the Atmospheric Dispersion Model [48], which uses the advection–diffusion equation to describe the transport of pollutants:

$$\frac{\partial C}{\partial t} + \nabla.(vC) = D\nabla^2 C + S$$

where C is the pollutant concentration, v is the wind velocity vector, D is the diffusion coefficient, S is the pollutant source term. Mechanistic models provide high interpretability and are valuable for understanding pollution dynamics, ecosystem interactions, and hydrological cycles. However, they require detailed input data (e.g., emission inventories, meteorological data) and can be computationally expensive. AI-based surrogate models can reduce computational costs by approximating mechanistic model outputs using machine learning.

Hybrid Models: Hybrid models combine empirical and mechanistic approaches with AI techniques to improve prediction accuracy, computational efficiency, and adaptability. These models leverage the strengths of both approaches, empirical models which excel in capturing patterns from historical data but lack physical interpretability, and mechanistic models which offer scientific rigor but can be computationally expensive and require extensive input data. A common hybrid modeling approach is data assimilation, where AI integrates real-time observations into mechanistic models to dynamically adjust parameters and improve forecasting accuracy. For example, hybrid air quality models use AI to refine predictions from traditional chemical transport models (CTMs) like WRF-Chem or CMAQ:

$$C_{corrected} = C_{CTM} + f_{AI}(X)$$

where $C_{corrected}$ is the AI-enhanced air quality prediction, C_{CTM} is the output from a chemical transport model, $f_{AI}(X)$ is a correction function based on real-time observations and AI algorithms. Hybrid models are widely used in climate prediction, hydrological forecasting, and pollution dispersion modeling [49, 50]. They provide a balanced approach, ensuring scientific robustness while leveraging AI's ability to handle large-scale, high-dimensional data. Each modeling approach empirical, mechanistic, and hybrid plays a crucial role in AI-driven environmental science. These models are essential for addressing global environmental challenges, from climate change prediction to pollution control and resource management. AI continues to push the boundaries of mathematical modeling, making environmental predictions more accurate, scalable, and actionable.

2.1.2 Why Mathematical Modeling Matters in AI-Driven Environmental Research

Mathematical modeling plays a fundamental role in AI-driven environmental research, providing structured methods for analyzing and predicting complex environmental phenomena. AI enhances traditional mathematical models by integrating large datasets, improving computational efficiency, and capturing intricate relationships that conventional models struggle to handle. Below are three key reasons why mathematical modeling is critical in AI-based environmental studies (as shown in Fig. 1).

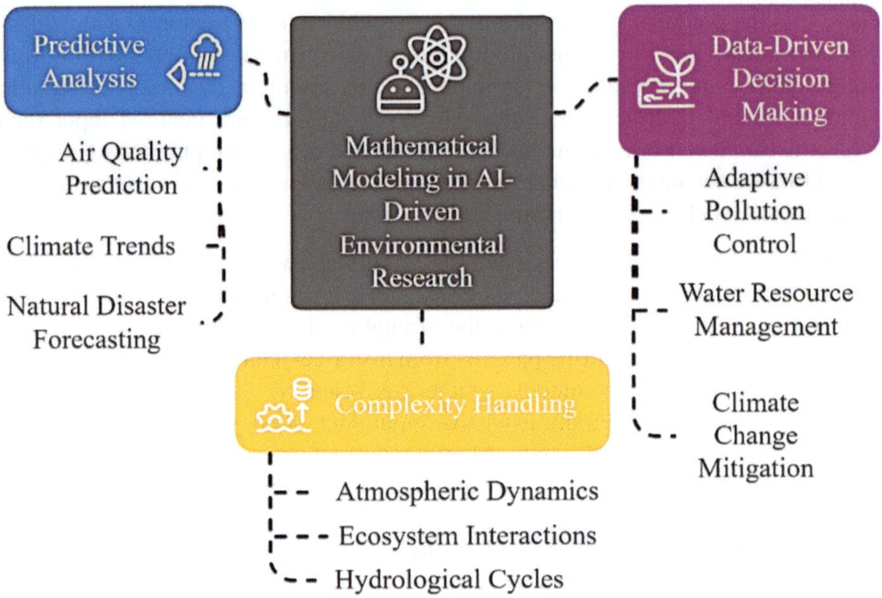

Fig. 1 AI-driven environmental research: role of mathematical modeling

Predictive Analysis: Predictive analysis is one of the most valuable applications of AI-integrated mathematical models in environmental science. Traditional forecasting methods rely on historical data and predefined equations, but AI-driven models adapt dynamically to new data inputs, improving prediction accuracy for pollutant levels, climate trends, and natural disasters [51]. AI-driven models analyze historical pollution data alongside real-time meteorological conditions to forecast air pollutant concentrations (e.g., NO_2, $PM_{2.5}$) in urban areas. This helps policymakers implement air quality control measures. AI models process large-scale satellite imagery, sea surface temperature records, and atmospheric CO_2 levels to project long-term climate patterns. Advanced deep learning techniques, such as recurrent neural networks (RNNs), help predict temperature changes, rainfall variability, and extreme weather events. AI-enhanced hydrological and geophysical models detect anomalies in seismic activity, ocean currents, and atmospheric pressure to predict earthquakes, hurricanes, and floods with higher accuracy than traditional statistical models [52]. Further chapters of this book also explain the application of AI in air quality prediction, climate trends, and natural disaster forecasting. By integrating AI with mathematical models, predictive analytics become faster, more precise, and adaptable to changing environmental conditions, allowing for proactive rather than reactive decision-making.

Data-Driven Decision Making: AI-driven mathematical models enable real-time, data-driven decision-making by integrating multiple sources of environmental information, such as: Satellite observations (e.g., NASA MODIS, Sentinel-5P for air pollution monitoring), IoT sensor networks (e.g., ground-based air and water quality monitoring stations), Historical environmental records (e.g., temperature trends, emission inventories). Unlike traditional models, which often rely on static equations and pre-set assumptions, AI-based models continuously learn from incoming data streams, refining their predictions and recommendations accordingly. For example, AI-enhanced models analyze real-time emissions data from industrial zones and adjust air quality policies dynamically [53]. If pollution levels exceed a threshold, AI models can suggest emission reduction strategies, such as traffic restrictions or industrial process modifications. AI-driven hydrological models process rainfall, groundwater levels, and river discharge rates to optimize water distribution and prevent shortages. These models help governments allocate water resources efficiently, reducing waste and ensuring sustainability. AI-enhanced economic and environmental models assess the long-term impact of different policy choices (e.g., carbon taxation, renewable energy incentives), allowing decision-makers to develop more effective climate action plans. By integrating big data analytics with AI-based mathematical models, environmental researchers can make faster, more informed decisions to mitigate risks and optimize resource allocation.

Complexity Handling: Environmental systems are inherently nonlinear, dynamic, and high-dimensional, making them difficult to model using traditional mathematical approaches. AI-driven models handle this complexity more effectively by learning intricate patterns and relationships between environmental variables. For example, the atmosphere is governed by highly nonlinear differential equations that describe turbulence, heat transfer, and moisturizing transport. AI-enhanced models can approximate these processes more efficiently than traditional physics-based models, leading to improved weather and climate predictions. Environmental models often need to simulate interactions between multiple species, pollutants, and climate factors. AI-based ecological models use graph neural networks (GNNs) and agent-based modeling to track complex species relationships, deforestation impacts, and habitat shifts [54, 55]. Traditional hydrological models struggle to simulate real-time flood risks, as they rely on historical water flow patterns. AI-driven models integrate remote sensing data, real-time precipitation monitoring, and terrain analysis to predict floods more accurately, helping authorities prepare for extreme weather events. By capturing high-dimensional dependencies and nonlinear interactions, AI-enhanced mathematical models improve our ability to simulate, analyze, and manage complex environmental systems.

2.1.3 How AI Enhances Traditional Mathematical Models for Prediction and Decision-Making

Traditional mathematical models often rely on simplified assumptions due to computational limitations and data scarcity. AI overcomes these limitations as summarized in Fig. 2.

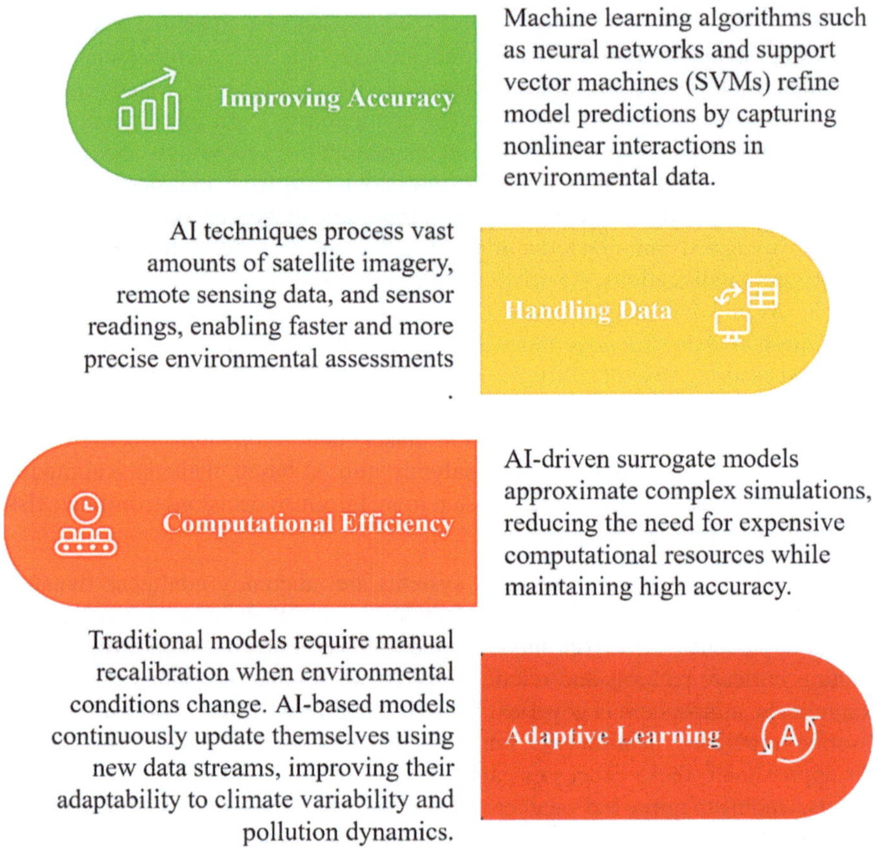

Fig. 2 Summary of key AI enhancements on traditional mathematical models for prediction and decision-making

Improving Accuracy: Machine learning algorithms such as neural networks and support vector machines (SVMs) are designed to capture complex, nonlinear relationships in data [56]. Consider the following formulation for a neural network, where the output is a function of the input data x and the learned parameters (weights) w:

$$y = f(x; w) = \sigma(W_2\sigma(W_1x + b_1) + b_2$$

where x is the input feature vector, W_1 and W_2 are the weights of the layers, b_1 and b_2 are the biases of the layers, σ is a nonlinear activation function, such as ReLU or sigmoid. By iterating and adjusting these weights during the training process using techniques like backpropagation, neural networks can learn complex nonlinear patterns in environmental data, thus enhancing accuracy in predictions.

Handling Large-Scale and Real-Time Data: AI techniques, such as deep learning and reinforcement learning, are designed to handle large datasets efficiently. For instance, a common technique to process large datasets (such as satellite imagery or sensor data) is convolutional neural networks (CNNs). The CNN for image classification can be described as:

$$y = CNN(X) = softmax(pool(conv(X, W)))$$

where, X is the input image or data matrix, W are the convolutional filters, *conv* represents the convolution operation, and pool is the pooling operation for dimensionality reduction. The output y represents class probabilities (in the case of classification tasks like environmental monitoring). By using techniques like data parallelism and distributed computing, AI can process vast amounts of real-time data quickly [57], allowing for faster environmental assessments.

Enhancing Computational Efficiency: AI-driven surrogate models, such as Gaussian processes (GPs), provide a computationally efficient way to approximate complex simulations. The surrogate model $f_{surrogate}$ can be trained on a limited set of high-fidelity simulation data $D = \{(x_i, y_i)\}$:

$$f_{surrogate}(x) = m(x) + K(x, x_i)K(x, x_i)^{-1}(y_i - m(x_i))$$

where, m(x) is the mean function, $K(x, x_i)$ is the covariance function (kernel), which defines the similarity between points, $K(x, x_i)^{-1}$ is the inverse of the covariance matrix, and y_i is the observed output for input x_i. By using a surrogate model like this, expensive high-fidelity simulations can be replaced with cheaper approximations, significantly enhancing computational efficiency.

Adaptive Learning and Model Updating: AI models, particularly online learning or sequential learning models, can continuously update as new data becomes available [58]. The update rule for a model parameter θ at time t is given by:

$$\theta_{t+1} = \theta_t + \eta \nabla_\theta \mathcal{L}(\theta_t; X_{t+1}, y_{t+1})$$

where, η is the learning rate, $\nabla_\theta \mathcal{L}(\theta_t; X_{t+1}, y_{t+1})$ is the gradient of the loss function $\mathcal{L}$ with respect to θ, computed for the new data point (X_{t+1}, y_{t+1}). This update process allows the model to adapt and learn from new environmental data, continuously improving its accuracy over time.

2.2 *Core Mathematical Concepts in AI for Environmental Science*

This subsection explores the mathematical concepts form the foundation of AI applications in environmental science. These concepts enable the development of models that optimize predictions, handle uncertainty, simulate dynamic systems,

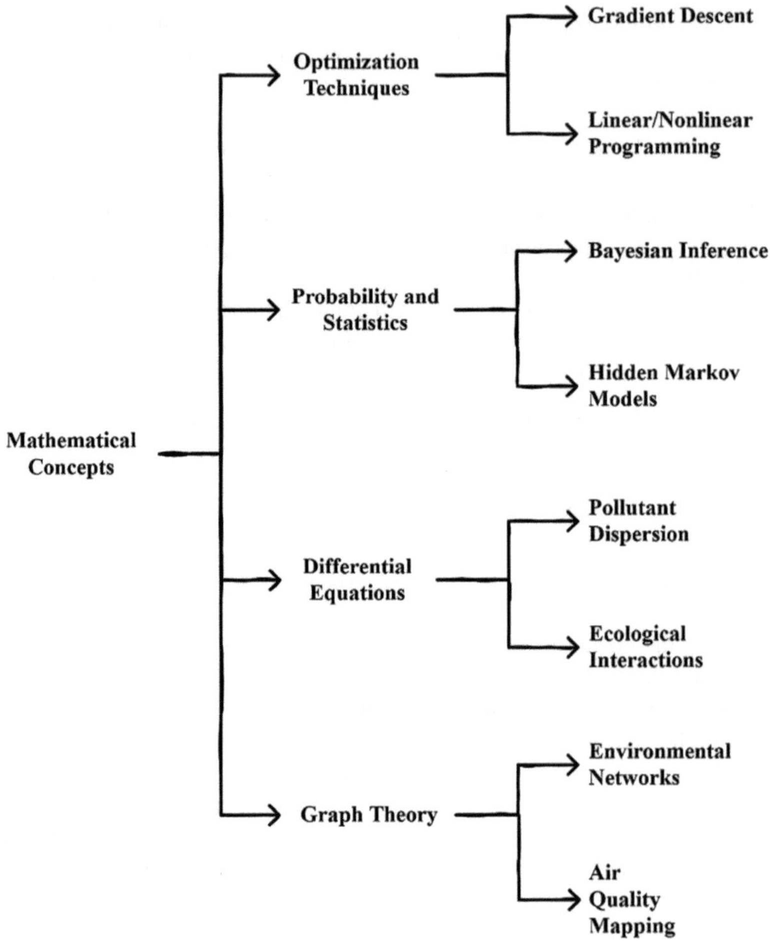

Fig. 3 Key mathematical principles that support AI-driven environmental modeling

and analyze complex networks. Below, we discuss key mathematical principles that support AI-driven environmental modeling (as shown in Fig. 3).

2.2.1 Optimization Techniques

Optimization plays a crucial role in training AI models for environmental applications [59]. Two fundamental optimization approaches include Gradient Descent and Linear and nonlinear programming. The Gradient descent is an iterative algorithm which minimizes the loss function in machine learning models by adjusting model parameters based on the gradient of the error. Given a function $f(\theta)$ representing the model's loss, the parameter update rule is:

$$\theta_{t+1} = \theta_t - \eta \nabla f(\theta_t)$$

where, η is the learning rate, and $\nabla f(\theta_t)$ is the gradient of the loss function. In environmental science, gradient descent is used in training deep learning models for pollution forecasting and climate simulations [60]. Linear and nonlinear programming is used as many environmental problems involve constrained optimization [61]. The linear programming is applied when the relationships between variables are linear:

$$\underset{x}{\min} c^T x, \text{ subject to } Ax \leq b$$

where x is the decision variable, c is the cost vector, and $Ax \leq b$ represents constraints. Nonlinear programming extends this approach to more complex relationships, such as optimizing air pollution control strategies under dynamic environmental conditions.

2.2.2 Probability and Statistics

Uncertainty is inherent in environmental systems, making probability and statistics essential for AI-based predictions. The Bayesian Inference comes into play, Bayesian methods update probability distributions as new data becomes available. Given prior knowledge $P(\theta)$ and observed data $P(D|\theta)$, the posterior probability is:

$$P(\theta|D) = \frac{P(D|\theta)P(\theta)}{P(D)}$$

Bayesian approaches are widely used in air quality modeling to integrate real-time sensor data and improve predictions [62]. Another method is named Hidden Markov Models (HMMs), the HMMs model time-series environmental data by assuming hidden states that evolve over time [63]. The probability of transitioning between states is governed by:

$$P(S_t|S_{t-1}) = A_{ij}, \ldots P(O_t|S_t) = B_{ik}$$

where A_{ij} represents transition probabilities and B_{ik} defines the observation probabilities. HMMs are used in atmospheric pollution tracking and weather pattern analysis [64].

2.2.3 Differential Equations

Environmental processes often involve continuous changes over time and space, which can be captured using differential equations. The advection–diffusion equation models the transport of pollutants in the atmosphere, the equation for this has already been given while defining mechanistic models in Sect. 2.1.1. This equation is crucial in AI-driven air quality modeling and dispersion forecasting [65]. Population dynamics and ecosystem changes are often modeled using Lotka-Volterra equations:

$$\frac{dN_1}{dt} = r_1 N_1 - \alpha_{12} N_1 N_2$$

$$\frac{dN_2}{dt} = r_2 N_2 - \alpha_{21} N_1 N_2$$

where N_1 and N_2 are population sizes, r_1 and r_2 are growth rates, α_{12} and α_{21} are interaction coefficients. These equations assist in AI-based ecological modeling and biodiversity assessments.

2.2.4 Graph Theory

Graph theory provides a powerful mathematical framework for modeling and analyzing environmental systems, particularly in pollution monitoring, climate networks, and ecological interactions [66]. A graph is a set of nodes (vertices) connected by edges that represent relationships between environmental factors. An environmental network can be represented as a graph:

$$G = (V, E)$$

where $V = \{v_1, v_2, \ldots, v_n\}$ is the set of vertices (nodes), each representing an entity such as pollution sources, monitoring stations, or regions in a geographical area. $E = \{e_{ij}\}$ is the set of edges, which define the relationships between entities, such as pollutant transport, spatial correlations, or meteorological dependencies. Each edge e_{ij} may have a weight w_{ij}, representing the strength or intensity of interaction between nodes v_i and v_j. Moreover, a graph is often represented using an adjacency matrix A, where:

$$A_{ij} = \begin{cases} w_{ij}, & \textit{if there is an edge between nodes i and j} \\ 0, & \textit{otherwise} \end{cases}$$

For unweighted graphs, $A_{ij} = 1$ if there is an edge and 0 otherwise. For weighted graphs, A_{ij} contains real values representing interaction strength (e.g., pollutant concentration correlation). The types of graphs used in environmental networks are Directed graphs (Digraphs), they are used when interactions are asymmetric. For example, wind-driven pollution transport is directional, where node v_i influences v_j but not vice versa. The Undirected graphs, used for symmetric interactions, such as regional air quality monitoring stations that share data without predefined directionality. The Weighted graphs where edges have numerical values, such as pollutant transport coefficients or similarity scores between monitoring locations. Dynamic graphs, used for time-evolving environmental systems, such as daily variations in air quality, where edges and weights change over time.

The spatial air quality data is represented using graph convolutional networks (GCNs), which extend traditional convolution operations to non-Euclidean spaces. The GCN update rule is:

$$H^{(l+1)} = \sigma\left(D^{-\frac{1}{2}}AD^{-\frac{1}{2}}H^{(l)}W^{(l)}\right)$$

where, $H^{(l)}$ is the feature matrix at layer l, A is the adjacency matrix, D is the degree matrix, $D^{-\frac{1}{2}}AD^{-\frac{1}{2}}$ normalizes the adjacency matrix, and $W^{(l)}$ are the learnable weights. GCNs enhance spatial air pollution analysis and forecasting.

These mathematical concepts like optimization, probability, differential equations, and graph theory are essential in AI-driven environmental modeling. They enable precise predictions, efficient computations, and robust analysis of complex environmental systems, supporting decision-making in pollution control, climate studies, and ecological assessments.

2.3 Machine Learning and Mathematical Formulations

Machine learning plays a crucial role in environmental modeling by leveraging mathematical techniques to analyze complex data patterns, predict pollutant concentrations, and improve decision-making. This section discusses key machine learning approaches and their mathematical formulations in environmental science.

2.3.1 Regression Models

Regression models establish relationships between environmental variables and pollutant concentrations [67]. They serve as fundamental tools for forecasting air pollution levels based on meteorological and emission data. The first basic ones are Linear regression models which sets up the relationship between a dependent variable Y (e.g., pollutant concentration) and independent variables $X_1, X_2, X_3, \ldots, X_n$ (e.g., temperature, wind speed, emissions):

$$Y = \beta_0 + \beta_1 X_1 + \beta_2 X_2 + \beta_3 X_3 + \cdots + \beta_n X_n + \epsilon$$

where, β_0 is the intercept, $\beta_1, \beta_2, \beta_3, \ldots \beta_n$ are regression coefficients, and ϵ represents the error term. The model parameters are estimated by minimizing the sum of squared errors (Ordinary Least Squares, OLS):

$$\min_{\beta} \sum_{i-1}^{m} (Y_i - \widehat{Y_i})^2$$

where, m is the number of observations. Linear regression is commonly used for air quality trend analysis. For more complex relationships, polynomial regression extends linear regression by incorporating higher-order terms:

$$Y = \beta_0 + \beta_1 X + \beta_2 X^2 + \beta_3 X^3 + \cdots + \beta_n X^n + \epsilon$$

where n is the polynomial degree. This is useful in modeling nonlinear pollutant dispersion trends influenced by meteorology. Moreover, the nonlinear models accommodate more complex interactions, such as exponential, logarithmic, and power-law relationships:

$$Y = \beta_0 e^{\beta_1 X} + \epsilon$$

These models are particularly useful in atmospheric chemistry, where pollutant concentrations exhibit exponential decay or growth patterns.

2.3.2 Support Vector Machines and Neural Networks

Support Vector Machines (SVMs) and Neural Networks are widely used in environmental monitoring due to their ability to capture complex relationships in high-dimensional data [68].

$$\min_{w,b} \frac{1}{2} w^2 \text{ subject to } \left| Y_i - \left(w^T X_i + b \right) \right| \leq \varepsilon$$

where w is the weight vector, b is the bias term, and ε is a predefined margin for acceptable error. By introducing kernel functions $K(X_i, X_j)$, SVMs transform data into higher-dimensional spaces to capture nonlinear relationships. Additionally, the Artificial Neural Networks (ANNs) consist of interconnected layers of neurons that learn patterns from environmental data [69]. A basic feedforward neural network is mathematically represented as:

$$Z^l = W^{(l)} X^{(l-1)} + b^l$$
$$A^{(l)} = f\left(Z^{(l)}\right)$$

where, $W^{(l)}$ and b^l are weight matrices and biases, and f(Z) is the activation function (e.g., ReLU, sigmoid). Training a neural network involves optimizing the loss function using gradient descent and backpropagation:

$$\frac{\partial L}{\partial W^{(l)}} = \frac{\partial L}{\partial A^{(l)}} \frac{\partial A^{(l)}}{\partial Z^{(l)}} \frac{\partial Z^{(l)}}{\partial W^{(l)}}$$

where, L is the loss function, $W^{(l)}$ is the weight of layer l, $A^{(l)}$ is the activation of layer l, $Z^{(l)}$ is the pre-activation (weighted sum before applying activation function) at layer l, $\frac{\partial L}{\partial W^{(l)}}$ is the gradient of loss with respect to weights (used for updating weights), $\frac{\partial L}{\partial A^{(l)}}$ is the gradient of loss with respect to activation, $\frac{\partial A^{(l)}}{\partial Z^{(l)}}$ is the gradient of activation with respect to pre-activation (derivative of activation function), and $\frac{\partial Z^{(l)}}{\partial W^{(l)}}$ is the gradient of pre-activation with respect to weights.

2.3.3 Ensemble Models

Ensemble learning improves model accuracy by combining multiple weak learners into a stronger predictor. Common ensemble techniques include stacking, boosting, and bagging [70]. Bagging (Bootstrap Aggregating) reduces variance by training multiple models on bootstrapped datasets and averaging predictions. A classic example is Random Forests, where each decision tree $h_i(X)$ is trained independently, and the final prediction is:

$$\widehat{Y} = \frac{1}{B}\sum_{i-1}^{B} h_i(X)$$

where, B is the number of trees. Bagging is useful in reducing overfitting in air pollution prediction models. Whereas in the Boosting technique, it sequentially trains models, with each iteration correcting errors from the previous one. Gradient Boosting Machines (GBMs) minimize the loss function iteratively:

$$F_m(X) = F_{m-1}(X) + \gamma h_m(X)$$

where, γ is the learning rate and $h_m(X)$ is the weak learner. XGBoost and LightGBM are popular boosting algorithms used for air quality forecasting [71]. Additionally in the Stacking technique, it combines multiple base models (e.g., SVMs, neural networks) by training a meta-model that learns how to optimally combine predictions:

$$\widehat{Y} = g(h_1(X), h_2(X), \ldots, h_n(X))$$

where is the meta-learner function. This approach is beneficial for integrating diverse machine learning models in pollution analysis.

Mathematical formulations underpin machine learning models in environmental science, from simple regression techniques to advanced deep learning architectures. Regression models capture pollutant trends, SVMs and neural networks handle nonlinearities, and ensemble learning improves prediction accuracy, making these techniques essential for AI-driven environmental monitoring.

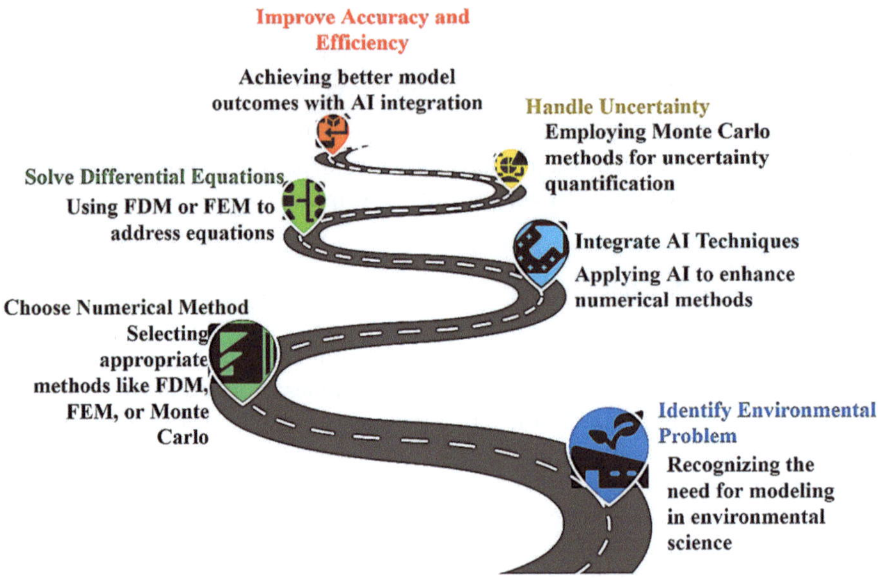

Fig. 4 Roadmap to get improved accuracy and efficiency using AI

2.4 *Numerical Methods and AI-Driven Simulations*

Mathematical modeling in environmental science often involves solving complex differential equations and handling uncertainty in predictions. Traditional numerical methods, such as the Finite Difference Method (FDM), Finite Element Method (FEM), and Monte Carlo simulations provide essential tools for these tasks. AI enhances these numerical approaches by improving accuracy, computational efficiency, and adaptability to real-world environmental conditions (as shown in Fig. 4).

2.4.1 Finite Difference and Finite Element Methods in AI-Based Environmental Simulations

The Finite Difference Method (FDM) is a numerical technique for solving differential equations by approximating derivatives using discrete points [72]. It is widely used in air quality modeling, pollutant dispersion, and fluid dynamics simulations. For a general function $u(x, t)$, its first-order derivative can be approximated using forward, backward, or central difference schemes:

Forward difference:

$$\frac{\partial u}{\partial x} \approx \frac{u(x + \Delta x) - u(X)}{\Delta x}$$

Backward difference:

$$\frac{\partial u}{\partial x} \approx \frac{u(x) - u(x - \Delta x)}{\Delta x}$$

Central Difference (more accurate):

$$\frac{\partial u}{\partial x} \approx \frac{u(x + \Delta x) - u(x - \Delta x)}{2\Delta x}$$

These approximations allow numerical simulations of pollution transport, atmospheric diffusion, and weather forecasting. The Finite Element Method (FEM) is another numerical approach used to solve partial differential equations (PDEs) by dividing a spatial domain into small, interconnected elements. Unlike FDM, which uses fixed grid points, FEM applies a mesh-based approach, making it ideal for handling irregular geometries in environmental modeling (e.g., pollutant dispersion in urban landscapes). A general PDE of pollution concentration $C(x, t)$:

$$\frac{\partial C}{\partial t} + u\frac{\partial C}{\partial x} = D\frac{\partial^2 C}{\partial x^2} + S$$

(where u is wind speed, D is diffusion coefficient, and S is a source term) can be discretized and solved using FEM with AI-enhanced adaptive mesh refinement techniques to optimize computation. AI improves these numerical methods by reducing computational complexity. Machine learning models, such as neural networks, can approximate numerical solutions, reducing the need for high-resolution simulations. AI-driven mesh refinement in FEM (adaptive meshing) improves accuracy by focusing computational effort on regions of high pollutant concentration. Additionally, AI uses real-time sensor data to update boundary conditions dynamically, improving environmental model predictions.

2.4.2 Monte Carlo Methods for Uncertainty Quantification in Pollution Modeling

The Monte Carlo Method is a statistical technique widely used to model uncertainty in complex environmental systems [73]. It relies on repeated random sampling to estimate probabilistic outcomes, making it particularly useful for assessing variability in environmental data. In pollution modeling, Monte Carlo simulations help evaluate uncertainties in emissions data [74], fluctuations in meteorological conditions, and the potential risks associated with public health impacts. By incorporating randomness into simulations, this method provides a more comprehensive understanding of the range of possible outcomes, allowing for more informed decision-making in environmental management. The Monte Carlo simulation process in pollution modeling begins by defining a model $C = f(x)$ that describes pollutant concentration based on environmental factors such as emissions, atmospheric conditions, and chemical transformations. Next, uncertain parameters, including emission rates, wind speed, and background pollution levels, are identified. The method then generates numerous random samples from probability distributions representing these uncertainties. Each set of sampled parameters is used to compute the corresponding model output, and the distribution of results is analyzed to quantify uncertainty. The final probability distribution of pollutant concentrations offers valuable insights for policymakers, helping them assess potential pollution scenarios and implement effective air quality management strategies.

Artificial intelligence enhances Monte Carlo methods by significantly improving computational efficiency and precision. Neural networks can be trained to learn from previous Monte Carlo simulations, enabling fast sampling and reducing the number of required iterations. Bayesian deep learning further refines uncertainty quantification by incorporating real-time sensor data, making predictions more adaptive to changing environmental conditions. Additionally, AI-driven surrogate modeling replaces computationally expensive simulations with highly efficient approximations, allowing faster and more scalable uncertainty analyses. By integrating AI with Monte Carlo methods, researchers can perform more accurate and real-time assessments of pollution risks [75], ultimately contributing to better environmental decision-making and public health protection.

3 Conclusion

This chapter has established the groundwork for understanding the transformative potential of AI within environmental science, emphasizing its role in addressing ecological challenges and promoting sustainable practices. AI's ability to process and analyze vast, complex datasets from diverse sources such as satellite imagery, sensor networks, and geospatial information enables comprehensive environmental monitoring, modelling, and prediction. The integration of AI methodologies marks a paradigm shift in environmental monitoring, enhancing data collection, analysis, and interpretation through advanced sensors, satellite imagery, and IoT devices. Machine learning techniques, including neural networks, decision trees, and clustering algorithms, empower predictive modelling, facilitating accurate forecasts of air quality, water contamination, biodiversity shifts, and natural disaster occurrences. AI-driven techniques, such as machine learning, neural networks, and ensemble modelling, significantly improve climate modelling and prediction by analyzing intricate climate datasets and simulating complex interactions within the earth system.

This leads to more reliable climate projections and the identification of potential tipping points and extreme events. The application of AI offers a transformative approach to air and water quality monitoring, enabling the precise modelling of pollution dynamics and the early detection of contamination sources through advanced sensor networks, satellite imagery, and IoT devices. Furthermore, AI's capabilities extend to ecosystem management, natural disaster prediction and management, waste management, renewable energy optimization, conservation, agriculture, and carbon footprint reduction, offering innovative solutions for sustainable stewardship of natural resources.

The chapter also underscores the crucial role of mathematical modelling in AI-driven environmental applications, providing structured frameworks for analyzing and predicting complex environmental phenomena. It explores how AI enhances traditional modelling techniques, such as empirical, mechanistic, and hybrid models, by integrating large datasets, improving computational efficiency, and capturing intricate relationships. Key mathematical concepts, including optimization techniques, probability and statistics, differential equations, and graph theory, are essential for developing AI models that optimize predictions, handle uncertainty, simulate dynamic systems, and analyze complex networks in environmental science. Machine learning approaches, such as regression models, SVMs, and neural networks, are vital for analyzing complex data patterns, predicting pollutant concentrations, and improving decision-making in environmental modelling. Ensemble learning techniques further enhance model accuracy by combining multiple weak learners into a stronger predictor. Numerical methods like the FDM, FEM, and Monte Carlo simulations are crucial for solving complex differential equations and quantifying uncertainty in environmental predictions. AI enhances these numerical approaches by improving accuracy, computational efficiency, and adaptability to real-world conditions.

References

1. Lin-Greenberg, E. (2020). *Allies and artificial intelligence: Obstacles to operations and decision-making (Spring 2020).* Texas National Security Review.
2. Seng, K. P., Ang, L. M., & Ngharamike, E. (2022). Artificial intelligence Internet of Things: A new paradigm of distributed sensor networks. *International Journal of Distributed Sensor Networks., 18*(3), 15501477211062836.
3. Andronie, M., Lăzăroiu, G., Iatagan, M., Uţă, C., Ştefănescu, R., & Cocoşatu, M. (2021). Artificial intelligence-based decision-making algorithms, internet of things sensing networks, and deep learning-assisted smart process management in cyber-physical production systems. *Electronics, 10*(20), 2497.
4. Ivić, M. (2019). Artificial intelligence and geospatial analysis in disaster management. *The International Archives of the Photogrammetry, Remote Sensing and Spatial Information Sciences, 42*, 161–166.
5. Hoang, T.-D., Ky, N. M., Thuong, N. T. N., Nhan, H. Q., & Ngan, N. V. C. (2022). Artificial intelligence in pollution control and management: Status and future prospects. *Artificial Intelligence and Environmental Sustainability: Challenges and Solutions in the Era of Industry, 4*, 23–43.

6. Méndez, M., Merayo, M. G., & Núñez, M. (2023). Machine learning algorithms to forecast air quality: A survey. *Artificial Intelligence Review, 56*(9), 1–36.
7. Tuia, D., Kellenberger, B., Beery, S., Costelloe, B. R., Zuffi, S., Risse, B., Burghardt, T., et al. (2022). Perspectives in machine learning for wildlife conservation. *Nature communications., 13*(1), 792.
8. Balogun, A.-L., Tella, A., Baloo, L., & Adebisi, N. (2021). A review of the inter-correlation of climate change, air pollution and urban sustainability using novel machine learning algorithms and spatial information science. *Urban Climate, 40*, 100989.
9. Ghobadi, F., & Kang, D. (2023). Application of machine learning in water resources management: A systematic literature review. *Water, 15*(4), 620.
10. Willard, J., Jia, X., Xu, S., Steinbach, M., & Kumar, V. (2022). Integrating scientific knowledge with machine learning for engineering and environmental systems. *ACM Computing Surveys, 55*(4), 1–37.
11. Ullah, Z., Al-Turjman, F., Mostarda, L., & Gagliardi, R. (2020). Applications of artificial intelligence and machine learning in smart cities. *Computer Communications, 154*, 313–323.
12. Ayturan, Y. A., Ayturan, Z. C., & Altun, H. O. (2018). Air pollution modelling with deep learning: A review. *International Journal of Environmental Pollution and Environmental Modelling, 1*(3), 58–62.
13. Ewuzie, U., Bolade, O. P., & Egbedina, A. O. (2022). Application of deep learning and machine learning methods in water quality modeling and prediction: A Review. In *Current trends and advances in computer-aided intelligent environmental data engineering* (pp. 185–218).
14. Chapman, M., Xu, L., Lapeyrolerie, M., & Boettiger, C. (2023). Bridging adaptive management and reinforcement learning for more robust decisions. *Philosophical Transactions of the Royal Society B, 378*(1881), 20220195.
15. Ogie, R. I., Rho, J. C., & Clarke, R. J. (2018). Artificial intelligence in disaster risk communication: A systematic literature review. In *2018 5th International Conference on Information and Communication Technologies for Disaster Management (ICT-DM)*. IEEE.
16. Irrgang, C., Boers, N., Sonnewald, M., Barnes, E. A., Kadow, C., Staneva, J., & Saynisch-Wagner, J. (2021). Will artificial intelligence supersede earth system and climate models? arXiv preprint arXiv:2101.09126
17. Kok, I., Guzel, M., & Ozdemir, S. (2021). Recent trends in air quality prediction: An artificial intelligence perspective. In *Intelligent environmental data monitoring for pollution management* (pp. 195–221). Elsevier.
18. Karatzas, K. (2010). Artificial intelligence applications in the atmospheric environment: Status and future trends. *Environmental Engineering and Management Journal, 9*, 171–180.
19. Li, X., Peng, L., Hu, Y., Shao, J., & Chi, T. (2016). Deep learning architecture for air quality predictions. *Environmental Science and Pollution Research, 23*, 22408–22417.
20. Hanoon, M. S., Ahmed, A. N., Fai, C. M., Birima, A. H., Razzaq, A., Sherif, M., El-Shafie, A., et al. (2021). Application of artificial intelligence models for modeling water quality in groundwater: Comprehensive review, evaluation and future trends. *Water, Air, & Soil Pollution., 232*, 1–41.
21. Li, C. (2020). Biodiversity assessment based on artificial intelligence and neural network algorithms. *Microprocessors and Microsystems, 79*, 103321.
22. Khatun, R., Talukdar, S., Pal, S., Saha, T. K., Mahato, S., Debanshi, S., & Mandal, I. (2021). Integrating remote sensing with swarm intelligence and artificial intelligence for modelling wetland habitat vulnerability in pursuance of damming. *Ecological Informatics., 64*, 101349.
23. Tan, L., Guo, J., Mohanarajah, S., & Zhou, K. (2021). Can we detect trends in natural disaster management with artificial intelligence? A review of modeling practices. *Natural Hazards, 107*, 2389–2417.
24. Guha, S., Jana, R. K., & Sanyal, M. K. (2022). Artificial neural network approaches for disaster management: A literature review (2010–2021). *International Journal of Disaster Risk Reduction, 103276*.
25. Sun, W., Bocchini, P., & Davison, B. D. (2020). Applications of artificial intelligence for disaster management. *Natural Hazards, 103*(3), 2631–2689.

26. Güçlü, D., & Dursun, Ş. (2010). Artificial neural network modelling of a large-scale wastewater treatment plant operation. *Bioprocess and Biosystems Engineering, 33*, 1051–1058.
27. Gupta, P. K., Shree, V., Hiremath, L., & Rajendran, S. (2019). The use of modern technology in smart waste management and recycling: Artificial intelligence and machine learning. In *Recent advances in computational intelligence* (pp. 173–188).
28. Gopal, L. C., Govindarajan, M., Kavipriya, M., Mahboob, S., Al-Ghanim, K. A., Virik, P., Shankar, V., et al. (2021). Optimization strategies for improved biogas production by recycling of waste through response surface methodology and artificial neural network: Sustainable energy perspective research. *Journal of King Saud University-Science., 33*(1), 101241.
29. Garcia Marquez, F. P., & Peinado Gonzalo, A. (2021). A comprehensive review of artificial intelligence and wind energy. In *Archives of computational methods in engineering* (pp. 1–24).
30. Ramasamy, P., Chandel, S., & Yadav, A. K. (2015). Wind speed prediction in the mountainous region of India using an artificial neural network model. *Renewable Energy, 80*, 338–347.
31. Gonzalez, L. F., Montes, G. A., Puig, E., Johnson, S., Mengersen, K., & Gaston, K. J. (2016). Unmanned aerial vehicles (UAVs) and artificial intelligence revolutionizing wildlife monitoring and conservation. *Sensors, 16*(1), 97.
32. Isabelle, D. A., & Westerlund, M. (2022). A review and categorization of artificial intelligence-based opportunities in wildlife, ocean and land conservation. *Sustainability., 14*(4), 1979.
33. Sharma, S., Sato, K., & Gautam, B. P. (2023). A methodological literature review of acoustic wildlife monitoring using artificial intelligence tools and techniques. *Sustainability, 15*(9), 7128.
34. Mukwazvure, A., & Magadza, T. B. (2014). *A survey on anti-poaching strategies.*
35. Bannerjee, G., Sarkar, U., Das, S., & Ghosh, I. (2018). Artificial intelligence in agriculture: A literature survey. *International Journal of Scientific Research in Computer Science Applications and Management Studies., 7*(3), 1–6.
36. Murase, H. (2000). Artificial intelligence in agriculture. *Computers and Electronics in Agriculture, 29*(1/2).
37. Eli-Chukwu, N. C. (2019). Applications of artificial intelligence in agriculture: A review. *Engineering, Technology & Applied Science Research, 9*(4).
38. Smith, M. J. (2018). Getting value from artificial intelligence in agriculture. *Animal Production Science, 60*(1), 46–54.
39. Mayfield, H. J., Smith, C., Gallagher, M., & Hockings, M. (2020). Considerations for selecting a machine learning technique for predicting deforestation. *Environmental Modelling & Software, 131*, 104741.
40. Mor, S., Madan, S., & Prasad, K. D. (2021). Artificial intelligence and carbon footprints: Roadmap for Indian agriculture. *Strategic Change, 30*(3), 269–280.
41. Dhar, P. (2020). The carbon impact of artificial intelligence. *Nature Machine Intelligence, 2*(8), 423–425.
42. Soofastaei, A., Aminossadati, S. M., Kizil, M. S., & Knights, P. (2016). *Reducing fuel consumption of haul trucks in surface mines using artificial intelligence models.*
43. Ahmad, T., Zhang, D., Huang, C., Zhang, H., Dai, N., Song, Y., & Chen, H. (2021). Artificial intelligence in sustainable energy industry: Status Quo, challenges and opportunities. *Journal of Cleaner Production, 289*, 125834.
44. Petcu, M. A., Ionescu-Feleaga, L., Ionescu, B. -Ş, & Moise, D.-F. (2023). A decade for the mathematics: Bibliometric analysis of mathematical modeling in economics, ecology, and environment. *Mathematics, 11*(2), 365.
45. Jacob, D. J., & Winner, D. A. (2009). Effect of climate change on air quality. *Atmospheric Environment, 43*(1), 51–63.
46. Chaudhuri, S., & Dutta, D. (2014). Mann-Kendall trend of pollutants, temperature and humidity over an urban station of India with forecast verification using different ARIMA models. *Environmental Monitoring and Assessment, 186*, 4719–4742.
47. Nawab, F., Abd Hamid, A. S., Ibrahim, A., Sopian, K., Fazlizan, A., & Fauzan, M. F. (2023). Solar irradiation prediction using empirical and artificial intelligence methods: A comparative review. *Heliyon, 9*(6).

48. Dou, Z., Liu, Z., Li, L., Zhou, H., Wang, Q., Zhang, J., & Chen, L. (2022). Atmospheric dispersion prediction of accidental release: A review. *Emergency Management Science and Technology, 2*(1), 1–20.
49. Russo, A., & Soares, A. O. (2014). Hybrid model for urban air pollution forecasting: A stochastic spatio-temporal approach. *Mathematical Geosciences, 46*, 75–93.
50. Beevers, S. D., Kitwiroon, N., Williams, M. L., Kelly, F. J., Ross Anderson, H., & Carslaw, D. C. (2013). Air pollution dispersion models for human exposure predictions in London. *Journal of Exposure Science & Environmental Epidemiology, 23*(6), 647–653.
51. Leena Sri, R., Vetriveeran, D., Sambandam, R. K., Jenefa, J., & Thangavel, K. (2025). *Transformative trends in AI for environmental monitoring: Challenges, applications. environmental monitoring using artificial intelligence* (pp. 1–28).
52. Sun, Z., Cristea, N., & Rivas, P. (2023). *Artificial intelligence in earth science: Best practices and fundamental challenges*. Elsevier.
53. Adegbite, A. O., Barrie, I., Osholake, S. F., Alesinloye, T., & Bello, A. B. (2024). *Artificial intelligence in climate change mitigation: A review of predictive modeling and data-driven solutions for reducing greenhouse gas emissions.*
54. Han, H., Liu, Z., Li, J., & Zeng, Z. (2024). Challenges in remote sensing based climate and crop monitoring: Navigating the complexities using AI. *Journal of Cloud Computing, 13*(1), 1–14.
55. Gupta, D., Kumar, P., Shukla, M. M., & Mahapatra, S. (2024). Enhancing climate change prediction and risk assessment with deep learning: Architectural approaches and data challenges. In *AI for climate change and environmental sustainability* (pp. 19–36). CRC Press.
56. Ghosh, S., Dasgupta, A., & Swetapadma, A. (2019). A study on support vector machine based linear and non-linear pattern classification. In *2019 International Conference on Intelligent Sustainable Systems (ICISS)*. IEEE.
57. Duan, S., Wang, D., Ren, J., Lyu, F., Zhang, Y., Wu, H., & Shen, X. (2022). Distributed artificial intelligence empowered by end-edge-cloud computing: A survey. *IEEE Communications Surveys & Tutorials, 25*(1), 591–624.
58. Bidaki, S. A., Mohammadkhah, A., Rezaee, K., Hassani, F., Eskandari, S., Salahi, M., & Ghassemi, M. M. (2025). *Online continual learning: A systematic literature review of approaches, challenges, and benchmarks*. arXiv preprint arXiv:2501.04897
59. Kingston, G. B., Dandy, G. C., & Maier, H. R. (2008). Review of artificial intelligence techniques and their applications to hydrological modeling and water resources management Part 2–optimization. *Water Resources Research Progress, 2*, 67–99.
60. Masih, A. (2019) Machine learning algorithms in air quality modeling. *Global Journal of Environmental Science & Management (GJESM), 5*(4).
61. Avriel, M. (2003). *Nonlinear programming: Analysis and methods*. Courier Corporation.
62. Nguyen, H. A., Ha, Q. P., Duc, H., Azzi, M., Jiang, N., Barthelemy, X., & Riley, M. (2023). Long short-term memory Bayesian neural network for air pollution forecast. *IEEE Access, 11*, 35710–35725.
63. Patterson, T. A., Basson, M., Bravington, M. V., & Gunn, J. S. (2009). Classifying movement behaviour in relation to environmental conditions using hidden Markov models. *Journal of Animal Ecology, 78*(6), 1113–1123.
64. Igbawua, T., Tyovenda, A. A., Sombo, T., & Echi, I. M. (2024). Determination of background PM2. In *5 pollution levels in Nigeria using Hidden Markov Model* (pp. 4–18). Reading Time.
65. Union, J. I. G. (2013). Advection diffusion equation models in near-surface geophysical and environmental sciences. *The Journal of Indian Geophysical Union, 17*, 117–127.
66. Ghosh, S., Mallick, A., Chowdhury, A., De Sarkar, K., & Mukherjee, J. (2024). Graph theory applications for advanced geospatial modelling and decision-making. *Applied Geomatics, 16*(4), 799–812.
67. Hoek, G., Beelen, R., De Hoogh, K., Vienneau, D., Gulliver, J., Fischer, P., & Briggs, D. (2008). A review of land-use regression models to assess spatial variation of outdoor air pollution. *Atmospheric Environment, 42*(33), 7561–7578.

68. Ali, S. (2019). *Support vector machine (SVM) aggregation modelling for spatio-temporal air pollution analysis.*
69. Han, K., & Wang, Y. (2021). A review of artificial neural network techniques for environmental issues prediction. *Journal of Thermal Analysis and Calorimetry, 145*(4), 2191–2207.
70. Odegua, R. (2019). An empirical study of ensemble techniques (bagging, boosting and stacking). In *Proceedings, Conference on Deep Learning. indabaXAt*
71. Ansari, A., & Quaff, A. R. (2025). Advanced machine learning techniques for precise hourly air quality index (AQI) prediction in Azamgarh, India. *International Journal of Environmental Research, 19*(1), 15.
72. Mazumder, S. (2015). *Numerical methods for partial differential equations: Finite difference and finite volume methods.* Academic Press.
73. Wang, X. E., Zhan, W., & Wang, S. (2020). Uncertain water environment carrying capacity simulation based on the Monte Carlo method–system dynamics model: A case study of Fushun City. *International Journal of Environmental Research and Public Health, 17*(16), 5860.
74. Arhami, M., Kamali, N., & Rajabi, M. M. (2013). Predicting hourly air pollutant levels using artificial neural networks coupled with uncertainty analysis by Monte Carlo simulations. *Environmental Science and Pollution Research, 20*, 4777–4789.
75. Tan, Y., Robinson, A. L., & Presto, A. A. (2014). Quantifying uncertainties in pollutant mapping studies using the Monte Carlo method. *Atmospheric Environment, 99*, 333–340.

Chapter 2
Review on Artificial Intelligence in the Environmental Monitoring

Balendra V. S. Chauhan, Ajitanshu Vedrtnam, Kevin P. Wyche, and Sneha Verma

Abstract In the contemporary landscape of environmental science, this comprehensive review investigates the paradigm shift induced by Artificial Intelligence (AI) in traditional monitoring methodologies. The conventional frameworks, beset by limitations in scalability, efficiency, and data processing, have necessitated a transformative response, and AI emerges as the vanguard of this evolution. By scrutinizing the inherent constraints of traditional techniques, the review articulates the imperative for embracing advanced AI-driven solutions. The primary focus is directed towards delineating the manifold applications of AI in environmental monitoring, underscoring its capacity to augment precision, velocity, and overall efficacy. From predictive analyses of air quality dynamics to nuanced evaluations of water and soil conditions, AI algorithms manifest unparalleled prowess in processing and interpreting heterogeneous datasets. The narrative is enriched with real-world case studies, exemplifying successful AI implementations, and accentuating its potential to redefine the contours of environmental science. However, amidst the promise and prospects lie challenges inherent to the seamless integration of AI into the fabric of environmental monitoring. The review dissects these challenges, providing strategic insights into overcoming impediments and catalyzing the widespread adoption of

A. Vedrtnam
Mineral and Energy Economy Institute, Polish Academy of Sciences, Kraków, Poland
e-mail: ajitanshu.m@invertis.org

B. V. S. Chauhan (✉) · K. P. Wyche
Centre for Earth Observation Science, University of Brighton, Brighton, United Kingdom
e-mail: b.chauhan@brighton.ac.uk

K. P. Wyche
e-mail: kevinwyche@googlemail.com

B. V. S. Chauhan · S. Verma
School of Engineering, Newcastle University, Newcastle Upon Tyne, United Kingdom
e-mail: sneha.verma@newcastle.ac.uk

A. Vedrtnam
Department of Mechanical Engineering, Invertis University, Bareilly 243001, India

Institute of Construction and Architecture, Slovak Academy of Science, Bratislava, Slovakia

A. Vedrtnam et al., *Prospects of Artificial Intelligence in the Environment*,
https://doi.org/10.1007/978-981-96-6863-2_2

AI-driven methodologies. As our global community grapples with imminent environmental challenges, this review maps out a trajectory that positions AI as an indispensable instrument in advancing our comprehension of ecosystems. It advocates for a proactive, data-driven approach to decision-making, delineating a roadmap that is poised to ameliorate environmental concerns and contribute substantively to the cultivation of a sustainable future.

Keywords Artificial Intelligence (AI) · Environment · Environmental monitoring · Climate modeling · Prediction · Ecosystem management · Natural disaster management · Renewable energy optimization · Waste management · Conservation · Wildlife protection

1 Introduction

In the realm of environmental science, the integration of Artificial Intelligence (AI) heralds a transformative epoch, representing a departure from the constraints that have historically encumbered traditional monitoring methodologies. This shift is underscored by the intrinsic limitations of conventional frameworks, marked by inefficiencies in scalability, processing vast datasets, and comprehending the complexities of environmental systems. Over the course of environmental monitoring's history, challenges in manual data collection and oversimplified models have hindered its ability to provide holistic insights into intricate ecosystems [1]. AI, with its capacity to emulate human intelligence and adapt through learning, emerges as a disruptive force poised to redefine the landscape of environmental monitoring [2, 3]. This paradigmatic shift is not merely a technological upgrade but a profound reevaluation of how we perceive and respond to environmental dynamics. Historical analysis illuminates the inadequacies of traditional approaches, setting the stage for the pivotal role that AI can play in transcending these limitations [4]. Embarking on this exploration, the historical narrative becomes a critical lens, elucidating the imperative for embracing AI-driven solutions to address the longstanding challenges and elevate the efficacy of environmental monitoring practices [5].

Throughout history, the field of environmental monitoring has confronted substantial challenges that impede its ability to offer nuanced insights into intricate ecosystems. The traditional approaches, often reliant on manual data collection and simplistic models, have demonstrated inadequacies in the face of mounting environmental challenges [6]. As the global landscape witnesses an escalation in environmental issues, the shortcomings of these conventional methods become increasingly apparent. One of the primary impediments lies in the inability to scale operations efficiently. Traditional monitoring, rooted in manual collection processes, is inherently limited by manpower and time constraints [7]. The magnitude of environmental data to be collected and processed demands a more sophisticated and streamlined approach, prompting a shift towards solutions that leverage the capabilities of AI [8].

Inefficiencies in processing vast and diverse datasets further compound the challenges faced by traditional environmental monitoring. The sheer volume and heterogeneity of data generated by ecosystems necessitate advanced computational techniques for meaningful analysis [9]. Here, AI proves to be a game-changer. Its ability to handle massive datasets with speed and precision aligns seamlessly with the demands of environmental monitoring [10]. Researchers have noted that AI algorithms excel in interpreting complex environmental data, facilitating a more nuanced understanding of ecological dynamics [11]. Figure 1 summarizes the machine learning applications in environmental ecology. By employing machine learning, these algorithms can discern patterns, correlations, and anomalies in data, enabling a level of analysis that surpasses the capabilities of manual and rule-based approaches [12]. The intricate nature of environmental systems poses another layer of complexity that traditional methodologies struggle to address adequately.

Ecosystems are characterized by dynamic interdependencies, non-linear relationships, and multifaceted variables that defy simplistic models.

AI, equipped with its capacity for adaptive learning and complexity handling, emerges as a potent tool to navigate this intricacy. Studies by environmental scientists have emphasized the potential of AI to model and simulate complex ecological processes, providing a more accurate representation of real-world conditions. This

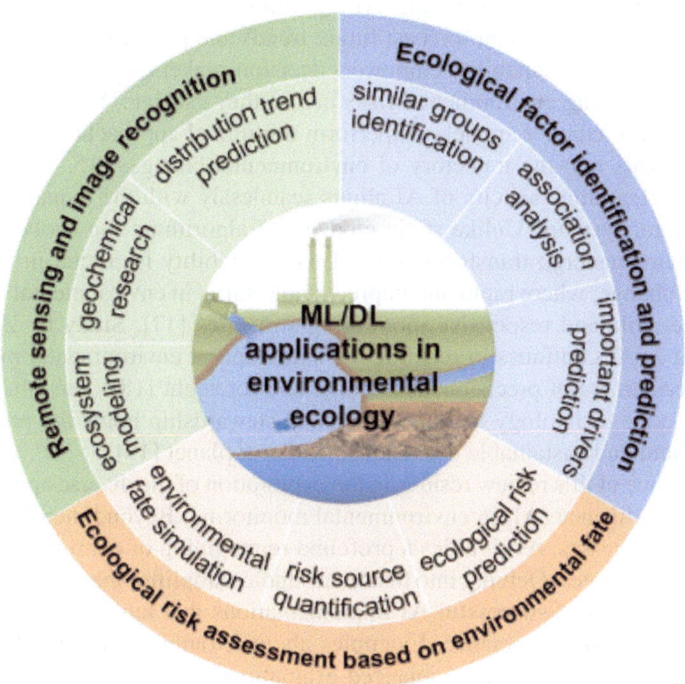

Fig. 1 Machine learning or deep learning applications in environmental ecology

departure from oversimplified models marks a paradigm shift in environmental monitoring, acknowledging the need for a more sophisticated approach to comprehend and respond to the multifaceted challenges posed by environmental changes. Against the backdrop of historical challenges in environmental monitoring, AI emerges as a transformative force, providing unparalleled capabilities to process, analyze, and interpret vast amounts of heterogeneous environmental data. The integration of AI into monitoring practices transcends mere technological intervention; it signifies a paradigmatic shift in our approach to perceiving, understanding, and responding to environmental dynamics. The historical trajectory of environmental monitoring underscores a persistent struggle to adapt to the intricacies of nature, paving the way for the disruptive potential of AI. At its core, AI revolutionizes environmental monitoring by harnessing advanced computational techniques, particularly machine learning algorithms, to unravel complex patterns within environmental datasets [13, 14]. Researchers have demonstrated that AI excels in handling the intricacies of diverse data sources, enabling a depth of analysis that was previously unattainable. Machine learning algorithms can discern subtle correlations, identify outliers, and reveal non-linear relationships within environmental data, providing scientists with a more holistic and accurate understanding of ecological systems [14].

The infusion of AI into environmental monitoring practices is not solely a response to the limitations of traditional methodologies; it represents a proactive stride towards predictive and prescriptive analytics. AI algorithms, trained on historical and real-time environmental data, can forecast future trends and potential ecological disruptions. This predictive capability empowers decision-makers to implement proactive measures, mitigating environmental risks before they escalate. Studies have shown that AI-driven predictive models outperform traditional approaches, offering more reliable insights into the trajectory of environmental changes [15, 16]. Moreover, the adaptive learning capacity of AI aligns seamlessly with the dynamic nature of environmental systems. Unlike static models, AI algorithms can evolve and adapt as ecosystems undergo transformations. This adaptability is crucial in the context of climate change, where rapid and unpredictable shifts in environmental conditions necessitate agile and responsive monitoring strategies [17]. Study by Zong et al. shows that as AI continues to redefine the landscape of environmental monitoring, it introduces an era of precision, scalability, and foresight [18]. The amalgamation of cutting-edge technology and environmental stewardship holds the promise of a more resilient and sustainable coexistence with our planet [19].

The essence of this review resides in the exploration of the diverse applications of Artificial Intelligence (AI) in environmental monitoring. Beyond merely enhancing established processes, AI signifies a profound reimagining of our approach to environmental challenges. Delving into historical studies amplifies the narrative, offering concrete instances of successful AI implementations that have reshaped the landscape of environmental science. Examples abound, ranging from predictive analyses of air quality dynamics to nuanced evaluations of water and soil conditions, and even real-time monitoring of biodiversity. These instances illuminate the transformative potential of AI across a spectrum of environmental domains. Machine learning algorithms, a cornerstone of AI, have demonstrated remarkable prowess in discerning complex patterns within environmental data. AI algorithms can process data from diverse sources, including satellite imagery, ground-based sensors, and meteorological observations, to provide real-time and predictive insights into air quality dynamics [13, 20–23]. These models can identify pollution sources, track the dispersion of pollutants, and forecast air quality changes with a level of accuracy that significantly surpasses conventional approaches [24, 25]. This not only enhances our understanding of the factors influencing air quality but also enables timely interventions to mitigate pollution and protect public health. In the context of water and soil monitoring, AI brings forth a revolution by enabling sophisticated analyses of complex interactions within ecosystems. Machine learning models can assimilate data on water quality parameters [26], soil composition [27, 28], and ecological indicators [29], offering comprehensive assessments of environmental health. This nuanced understanding is crucial for sustainable resource management [30], biodiversity conservation [31], and informed decision-making in agriculture and land use planning [32].

Antonelli et al. studied that the real-time monitoring of biodiversity is another domain where AI showcases its transformative potential [33]. Machine learning algorithms can process vast datasets of species distribution, behavior patterns, and ecological interactions, providing ecologists with valuable insights into the dynamics of ecosystems. These insights contribute to the development of effective conservation strategies, helping to safeguard biodiversity in the face of environmental changes. In essence, the multifaceted applications of AI in environmental monitoring mark a pivotal juncture in our scientific endeavors. Promises of progress, however, are tempered by challenges intrinsic to the seamless integration of AI into environmental monitoring frameworks, the challenges are discussed in Sect. 5 of this chapter. Ethical considerations, data security issues, and challenges related to interpretability emerge as substantial hurdles. The review revisits these challenges, providing strategic insights into overcoming impediments and catalyzing the widespread adoption of AI-driven methodologies. As our global community confronts pressing environmental challenges, the review's narrative unfolds a trajectory positioning AI as an indispensable instrument in advancing our comprehension of ecosystems. Further sections advocate for a proactive, data-driven approach to decision-making, it delineates a roadmap poised to ameliorate environmental concerns and substantively contribute to the cultivation of a sustainable. This comprehensive exploration integrates insights from historical studies on AI in environmental conservation, showcasing how advanced technologies can empower conservation efforts, enhance biodiversity monitoring, and contribute to ecological restoration. Synthesizing findings from the past, the review sheds light on the trajectory of advancements, offering a nuanced understanding of the evolution of AI applications in environmental contexts.

In essence, these expansive narrative weaves together historical perspectives, current challenges, and future trajectories, offering a panoramic view of the evolving relationship between AI and environmental monitoring. By synthesizing insights from past studies, the review serves as a comprehensive resource for researchers, policymakers, and practitioners, providing a historical lens that informs the challenges and opportunities at the intersection of AI and environmental science.

2 Overview of Environmental Monitoring

Environmental monitoring has undergone a transformative journey, evolving from basic observational practices to more advanced, technology-driven approaches. Traditionally, the focus was on manual data collection and simplistic models to assess air and water quality, soil conditions, biodiversity, and climate patterns. These methods, while foundational, often fall short in scalability, efficiency, and processing vast datasets. The subsequent sections will delve into the historical roots of environmental monitoring, tracing the trajectory from traditional approaches to the contemporary integration of Artificial Intelligence (AI). This exploration aims to illuminate the strengths and limitations of each phase, providing a comprehensive perspective on the evolution of monitoring methodologies.

2.1 Traditional Methods of Environmental Monitoring and Their Limitations

Traditional methods of environmental monitoring involve conventional techniques that have been historically employed to assess various aspects of the environment. These methods are often characterized by manual data collection and simple models. Here are some key aspects of traditional environmental monitoring:

2.1.1 Manual Data Collection

Historically, environmental monitoring heavily relied on manual data collection processes, representing a fundamental approach to understanding and evaluating environmental parameters [34]. This labor-intensive method involved deploying personnel to specific locations equipped with instruments designed to measure key factors such as air quality, water conditions, soil composition, and biodiversity. While providing valuable insights into local ecosystems, manual data collection faced inherent limitations. The spatial and temporal coverage of observations were often constrained, as deploying personnel to a multitude of locations regularly and frequently proved logistically challenging [34]. Consequently, this restricted the comprehensive understanding of environmental dynamics, hindering efforts to formulate effective conservation and management strategies.

Researchers have extensively scrutinized the shortcomings of manual data collection in environmental monitoring. Studies have demonstrated the limited scalability of this approach, particularly in the context of assessing air quality over large geographic areas [35]. Traditional methods struggled to capture real-time variations in air quality, especially during dynamic events such as industrial emissions or natural phenomena. Additionally, the delay between data collection and analysis posed challenges in responding promptly to emerging environmental threats. Recognizing these limitations, scientists explored alternative methodologies, seeking innovative solutions to enhance the efficiency and scope of environmental monitoring practices [36]. The subsequent evolution of monitoring techniques, including the integration of advanced technologies like remote sensing and AI-driven analytics, reflects a concerted effort to overcome the constraints of manual data collection [19].

2.1.2 Simplistic Models

Traditional monitoring methods often employed simplistic models to interpret environmental data, reflecting the computational constraints prevalent in earlier times [37]. These models aimed to simplify the intricate interactions within natural systems to facilitate analysis and prediction. However, their application faced significant challenges due to the inherent complexity of ecological processes. For instance, in air quality assessment, simplistic models struggled to account for the diverse sources of pollutants, their dispersion patterns, and the dynamic atmospheric conditions influencing their movement. This limitation led to less accurate predictions, hindering the formulation of targeted strategies for pollution control and mitigation.

Researchers and environmental scientists have extensively studied the drawbacks of simplistic models in environmental monitoring [38]. Past studies have highlighted instances where such models failed to adequately represent the complexities of aquatic ecosystems, impacting the precision of water quality assessments. The oversimplification of ecological dynamics posed challenges in predicting the impacts of human activities, such as industrial discharges or agricultural runoff, on freshwater ecosystems. Recognizing these shortcomings, the scientific community has increasingly emphasized the integration of more sophisticated modeling techniques, such as numerical simulations and machine learning algorithms, to capture the intricacies of environmental processes. These advancements aim to enhance the accuracy and reliability of environmental assessments, providing a more robust foundation for informed decision-making in resource management and conservation [39].

2.1.3 Limited Scalability

One of the significant drawbacks of traditional environmental monitoring methods lies in their limited scalability. Manual data collection and analysis processes proved to be time-consuming and resource-intensive, posing challenges when attempting to scale up monitoring efforts for broader geographic coverage or to address global environmental issues comprehensively. Researchers and environmental scientists often found themselves constrained by the constraints of traditional methodologies, impeding the ability to gather data efficiently across diverse ecosystems. For example, study conducted by Fascista, highlighted the challenges of implementing large-scale monitoring using traditional approaches [40]. These limitations spurred the exploration of innovative solutions such as AI to enhance scalability by automating data collection, analysis, and interpretation processes, allowing for more extensive coverage and a more holistic understanding of environmental dynamics.

In addition to this, the incorporation of AI-driven technologies in environmental monitoring has provided opportunities to overcome scalability challenges [41]. Advanced algorithms and machine learning models can handle vast datasets more efficiently, enabling monitoring initiatives to extend their reach beyond localized efforts. The work of Popescu et al. has showcased the potential for scalable environmental monitoring through AI, revolutionizing how we collect and process data on a larger scale [42]. As we navigate the era of AI-driven environmental monitoring, scalability emerges as a key benefit, fostering the ability to tackle global environmental challenges with a more comprehensive and expansive approach.

2.1.4 Inefficiencies in Processing Data

The processing of vast and diverse datasets posed significant challenges for traditional environmental monitoring methods. Inefficient data processing hindered the ability to derive timely and meaningful insights from the collected information. Traditional approaches struggled to keep pace with the increasing volume of environmental data, leading to delays in obtaining valuable information for decision-making. Studies conducted by Gilbert et al. underscored the inefficiencies in data processing within traditional methodologies, revealing a critical bottleneck in the workflow of environmental monitoring [43]. Recognizing these inefficiencies prompted a paradigm shift toward embracing AI-driven technologies to streamline data processing workflows. AI, with its capacity for rapid and automated analysis, has demonstrated effectiveness in overcoming the data processing challenges faced by traditional methods.

Furthermore, the integration of AI has introduced innovative techniques such as pattern recognition, anomaly detection, and predictive modeling to enhance data processing efficiency in environmental monitoring. The work of Konya and Nematzadeh exemplifies how AI applications have revolutionized the processing of environmental data, providing quicker and more accurate results [44]. As we transition from conventional to AI-enhanced methodologies, the focus on efficient data processing becomes a hallmark of contemporary environmental monitoring practices, enabling faster response times and more informed decision-making processes.

2.1.5 Reactiveness Instead of Proactiveness

A notable limitation of traditional monitoring approaches lies in their reactive nature, often responding to environmental issues as they arise rather than proactively anticipating and preventing problems. This reactive stance made it challenging to implement timely interventions and preventive measures, leaving ecosystems vulnerable to the consequences of delayed responses. Studies conducted by Sani et al. highlighted the limitations of conventional monitoring approaches in addressing environmental challenges, particularly due to their inherent response latency [45]. The inability to foresee and address emerging challenges in a timely manner undermined the effectiveness of traditional monitoring in maintaining environmental health.

The transition to AI-driven environmental monitoring addresses this limitation by introducing proactive capabilities. AI models, through continuous analysis of real-time data, enable the identification of potential issues before they escalate. Predictive analytics and early warning systems powered by AI empower environmental scientists to take preventive actions, mitigating the impact of disturbances. The work of Holloway et al. (2018) highlights how AI's proactive approach enhances environmental monitoring, transforming it from a reactive to a preventive discipline [46]. As we embrace AI technologies, the shift towards proactiveness becomes a cornerstone in reshaping environmental monitoring strategies for a more sustainable and resilient future, as further emphasized by the research of Hino et al., who demonstrate the potential of machine learning in predicting environmental risks and informing policy decisions [47].

2.1.6 Limited Real-Time Capabilities

A critical drawback of many traditional monitoring techniques lies in their limited real-time capabilities. These methods often involve manual data collection and time-consuming processes for analysis, making it challenging to respond promptly to dynamic environmental changes [45]. Real-time monitoring is imperative, especially in situations where rapid interventions are required to address emerging issues such as pollution events, climate fluctuations, or ecological disturbances. AI revolutionizes the landscape by providing advanced analytics and machine learning algorithms that facilitate real-time data processing. The integration of AI technologies enables environmental monitoring systems to detect and respond to changes almost instantaneously. This shift from delayed, reactive approaches to proactive, real-time monitoring enhances the agility of environmental management strategies, ensuring timely interventions and contributing to the overall effectiveness of environmental stewardship [48]. The implementation of real-time sensor networks, coupled with AI-driven analytics platforms, has significantly improved our ability to monitor and respond to environmental changes. These systems can continuously stream data, allowing for immediate analysis and anomaly detection. For instance, AI-powered air quality monitoring systems can provide instant feedback on pollution levels, enabling quick action to mitigate potential health risks and environmental damage [49].

Furthermore, AI's predictive capabilities enhance the proactive nature of environmental monitoring. By analyzing historical data and current conditions, AI models can forecast potential environmental issues before they occur. This predictive power allows authorities and environmental managers to implement preventive measures, rather than merely reacting to problems as they arise. For example, AI has been used to predict smog levels in cities, enabling timely interventions to protect public health [50]. The real-time capabilities of AI-driven environmental monitoring systems also facilitate improved accountability and compliance with environmental regulations. The ability to conduct continuous monitoring in remote areas ensures adherence to government regulations and corporate environmental commitments. Real-time data allows agencies and enterprises to set and track environmental performance goals, creating a more transparent and responsive approach to environmental management [51]. By enabling instant data processing, analysis, and prediction, AI transforms environmental monitoring from a reactive to a proactive discipline, significantly enhancing our ability to protect and manage our environment effectively.

3 Role of Artificial Intelligence in Environmental Monitoring

Artificial Intelligence (AI) has emerged as a powerful tool in environmental monitoring, offering unprecedented capabilities in data collection, analysis, and decision-making. AI technologies, including machine learning algorithms and deep neural networks, are being increasingly deployed to address complex environmental challenges and enhance our understanding of ecological systems [47, 52]. AI-driven environmental monitoring systems can process vast amounts of data from various sources, including satellite imagery, sensor networks, and citizen science initiatives. These systems can detect subtle changes in environmental parameters, predict future trends, and provide real-time insights for informed decision-making [53]. For instance, AI algorithms can analyze satellite images to track deforestation, monitor air and water quality, and assess the health of ecosystems on a global scale [54]. One of the key advantages of AI in environmental monitoring is its ability to handle large, complex datasets and identify patterns that may be difficult for human observers to detect. This capability allows for more accurate and timely detection of environmental issues, enabling faster response times and more effective conservation efforts [55].

3.1 Transformative Impact of AI on Environmental Surveillance

The integration of AI into environmental surveillance has revolutionized the way we monitor and protect our ecosystems. AI-powered systems have significantly enhanced the accuracy, efficiency, and scope of environmental monitoring efforts [56]. One of the most significant impacts of AI on environmental surveillance is the ability to conduct real-time monitoring and analysis. AI algorithms can process data from various sensors and sources instantaneously, providing up-to-date information on environmental conditions. This capability is particularly crucial for detecting and responding to rapid environmental changes or potential threats [57]. AI has also transformed remote sensing applications in environmental monitoring. Machine learning algorithms can analyze satellite imagery and aerial photographs to detect changes in land use, track wildlife populations, and monitor the spread of invasive species.

These applications have greatly expanded our ability to monitor large and remote areas that were previously difficult to access [6].

Furthermore, AI has enabled the development of predictive models that can forecast environmental changes and potential risks. By analyzing historical data and current trends, these models can predict future scenarios, allowing for proactive environmental management and conservation strategies [58]. The integration of AI in environmental monitoring has revolutionized our ability to predict and manage environmental changes. Advanced machine learning algorithms can analyse vast amounts of historical and real-time data to forecast future environmental scenarios with unprecedented accuracy. These predictive models enable proactive environmental management and conservation strategies, allowing stakeholders to anticipate and mitigate potential risks before they materialize [52]. In the field of environmental acoustics, AI has opened new avenues for biodiversity monitoring. AI algorithms can efficiently process and analyze audio recordings from natural habitats, identifying and tracking species based on their vocalizations. This approach provides valuable insights into ecosystem health and biodiversity trends, offering a non-invasive and scalable method for ecological research [59]. AI-driven systems have significantly enhanced pollution monitoring capabilities across air, water, and soil. These systems can detect and track various pollutants with high sensitivity, providing early warnings of pollution events and helping to identify pollution sources. This improved monitoring enables more effective and timely mitigation strategies, crucial for protecting both environmental and public health [60]. The transformative impact of AI extends to citizen science initiatives, democratizing environmental monitoring. AI-powered mobile applications and platforms have made it easier for citizens to contribute meaningful data to environmental monitoring efforts. This expansion of data collection scope and scale not only enhances the overall monitoring capabilities but also fosters public engagement in environmental issues [61].

AI has also revolutionized remote sensing applications for environmental monitoring. Machine learning algorithms, particularly deep learning models, can analyze satellite imagery and aerial photographs with unprecedented accuracy and speed. These systems can detect subtle changes in land use, vegetation cover, and ecosystem health across vast areas, providing crucial data for global environmental monitoring and conservation efforts [53]. In the realm of climate change research, AI has enhanced our ability to model complex climate systems and predict future scenarios. Deep learning algorithms can process and analyze climate data from diverse sources, including weather stations, satellites, and ocean buoys, to create more accurate and detailed climate models. These AI-driven models can simulate intricate interactions between different components of the Earth system, leading to improved predictions of climate change impacts and more effective adaptation strategies. The integration of AI with Internet of Things (IoT) technologies has led to the development of smart environmental monitoring networks [62–64]. These systems combine low-cost sensors, wireless communication technologies, and AI algorithms to create dense, real-time monitoring networks for air quality, water quality, and ecosystem health. For instance, smart buoys equipped with AI can analyze water samples in situ, detecting pollutants, algal blooms, and changes in marine ecosystems without the need for manual sample collection and laboratory analysis [65, 66]. Lastly, AI is playing an increasingly important role in wildlife conservation efforts [67]. Computer vision algorithms can analyze camera trap images and videos to identify and track individual animals, estimate population sizes, and monitor species behavior. This technology has been particularly valuable in studying elusive or endangered species in remote areas. Additionally, AI-powered drones equipped with thermal imaging cameras can detect poachers and illegal logging activities in protected areas, enhancing conservation efforts and law enforcement [68, 69].

3.2 Advanced Data Analysis and Prediction Techniques in AI for Environmental Monitoring

The proliferation of large-scale environmental monitoring systems has generated an unprecedented amount of data, prompting the need for advanced analytical tools to extract actionable insights in real-time. Artificial Intelligence (AI) has emerged as a transformative technology, capable of efficiently handling the volume, variety, and velocity of environmental datasets. Researchers have explored various AI-driven techniques to address challenges in data processing, pattern recognition, and predictive analytics, which are essential for making timely decisions in environmental management and public health interventions. The increasing availability of high-dimensional and heterogeneous environmental data has led to the development of advanced AI-based data analysis techniques. These methods go beyond traditional statistical and rule-based approaches, enabling more accurate, scalable, and efficient analysis of environmental monitoring datasets. Some of the most prominent advanced techniques used in the field:

3.2.1 Deep Learning (DL) Architectures

Deep learning, particularly convolutional neural networks (CNNs) and recurrent neural networks (RNNs), is widely used for analyzing spatiotemporal data in environmental monitoring [70]. CNNs are well-suited for processing satellite imagery, climate patterns, and pollution maps, while RNNs and Long Short-Term Memory (LSTM) networks are effective for time series forecasting, such as predicting air quality, temperature, and pollutant dispersion. These models can learn complex representations of environmental factors, making them particularly useful for handling vast amounts of unstructured data, such as sensor networks and satellite data streams.

The paper by Zhao and Zettsu explores the use of deep learning architectures, specifically Convolution Recurrent Neural Networks (CRNNs), for short-term prediction of atmospheric sensing data [70]. This approach combines Convolutional Neural Networks (CNNs) and Long Short-Term Memory (LSTM) networks to handle both spatial and temporal data, enhancing the accuracy of environmental monitoring. CNNs are used to extract spatial features from atmospheric sensing data. By processing data from multiple sensors, CNNs transform spatial inputs into feature maps, capturing essential spatial characteristics of environmental data. LSTMs are employed to model temporal dependencies in the data. This is crucial for capturing long-term temporal patterns in environmental variables, such as PM2.5 concentrations. The integration of CNNs and LSTMs into a CRNN framework allows for simultaneous spatial and temporal predictions. This hybrid model effectively processes complex space–time series data, improving the prediction of PM2.5 levels. Deep neural networks, particularly attention-based models, have proven effective in forecasting and reconstructing environmental data. In Yang's 2023 study, an attention-based ConvLSTM model was developed to enhance multi-step image series forecasting by capturing long-range dependencies within latent variables [71]. This approach addresses the limitations of traditional ConvLSTM networks, which struggle to extract global interactions among hidden features. Additionally, an attention-based deep residual neural network was proposed to optimize sensor placement, significantly improving computational efficiency. The methodologies were evaluated using real-world datasets, demonstrating a 7% reduction in mean squared error compared to existing methods and a 73% improvement in speed for sensor placement optimization, achieving approximately 90% accuracy. These advancements highlight the potential of deep neural networks in enhancing the precision and efficiency of environmental monitoring and assessment.

3.2.2 Transfer Learning

Transfer learning has emerged as a key method in environmental monitoring where labeled data is scarce or expensive to obtain. It allows pre-trained models (typically on a different but related task) to be fine-tuned on specific environmental datasets. For example, models trained on global climate data can be adapted to specific regions or pollution patterns, thereby reducing the need for extensive retraining and providing faster, more accurate insights. Transfer learning is being utilized in environmental monitoring to enhance the accuracy of soil heavy metal content retrieval in mining areas. In Yang et al.'s 2023 study, a Tr-GA-BP network was developed, leveraging pre-trained models with data from Landsat8 satellite images, digital elevation models, and geochemical samples [72]. This approach allows for effective mapping of copper and lead distributions with fewer samples, significantly reducing root mean square error (RMSE) and mean relative error (MRE) compared to traditional methods. The model's ability to adapt learned parameters from earlier datasets improves estimation precision and efficiency, demonstrating its potential in monitoring environmental pollutants.

3.2.3 Hybrid AI Models

Hybrid models combine traditional machine learning algorithms with AI techniques such as deep learning. One approach is to use ML methods like decision trees or random forests for feature extraction, followed by deep learning models for more refined predictions. Hybrid AI models, which combine multiple machine learning techniques, have shown great potential in enhancing environmental monitoring capabilities. These models leverage the strengths of different AI approaches to provide more accurate, robust, and comprehensive solutions for various environmental challenges. By integrating multiple AI techniques, hybrid models can better adapt to changing environmental conditions and handle diverse data types. This adaptability is crucial for real-time monitoring systems that need to adjust to dynamic environmental factors [73]. Hybrid AI models can process and analyze data from various sources simultaneously, providing a more holistic view of environmental conditions. For instance, they can integrate satellite imagery, sensor data, and historical records to assess soil quality or monitor deforestation [74]. Hybrid models combining LSTM networks with other techniques have shown high accuracy in predicting pollutant concentrations, such as PM10, in urban areas [73, 75]. For instance, Wen et al. proposed a novel spatiotemporal convolutional long short-term neural network that significantly improved air pollution prediction accuracy [76]. Water resource management has benefited significantly from hybrid AI approaches, with LSTM networks enhancing Internet of Things (IoT) systems for combined sewer overflow monitoring [77]. For soil quality assessment, AI-powered sensors and devices utilizing hybrid models can detect and quantify hazardous materials in soil and plants in real-time, ensuring food crop safety and preventing environmental contamination [78]. Hybrid AI models also play a crucial role in climate change modeling, integrating diverse data sources to support the understanding of climate change and its potential impacts [79]. In ecosystem health monitoring, these models can process vast amounts of satellite imagery to classify vegetation cover types, aiding in the identification of desertification trends and informing environmental planning. While hybrid AI models offer significant advantages, they often require more computational resources and careful design to balance the strengths of different AI techniques. Additionally, ensuring data quality and addressing potential biases remain crucial challenges in AI applications. Despite these considerations, the integration of hybrid AI models in environmental monitoring systems shows great promise in enhancing our ability to understand, predict, and manage environmental challenges more effectively.

3.2.4 Reinforcement Learning (RL)

Reinforcement learning (RL) has emerged as a powerful tool for optimizing environmental monitoring systems in dynamic and complex environments. RL algorithms can learn to make decisions, such as adjusting sensor networks or optimizing resource allocation for real-time data collection, allowing for adaptive and efficient monitoring strategies that can respond to changing environmental conditions and priorities. In the field of water resource management, RL has shown promising applications. For instance, Castelletti et al. proposed a novel approach using RL for multi-objective reservoir management [80]. Their study applied batch RL to design operating policies for water reservoirs, demonstrating the ability to learn from historical data and make real-time decisions. The approach offered an effective method for analyzing complex water management scenarios while optimizing multiple objectives simultaneously. RL has also been applied to air quality monitoring and prediction. Ayus et al. conducted a comprehensive comparison of machine learning and deep learning techniques for air pollution prediction, using China as a case study [81]. Their research highlighted the effectiveness of these advanced computational methods in forecasting Air Quality Index (AQI) and understanding annual pollution trends across various Chinese cities. This study demonstrates the growing importance of sophisticated AI techniques in environmental monitoring and their potential to inform practical action plans for controlling air pollution. In the realm of ecosystem monitoring, RL is being used to optimize the deployment and operation of autonomous environmental monitoring systems. Kober et al. discussed the challenges of applying RL techniques to real-world robotics problems, including environmental monitoring, and proposed innovative solutions to overcome them [82]. They focused on policy search methods in RL and their applications in robotics, which can be extended to autonomous environmental monitoring systems. These applications of RL in environmental monitoring not only enhance our understanding of environmental processes but also support evidence-based policymaking for environmental conservation. As the field continues to evolve, RL is expected to play an increasingly important role in addressing complex environmental challenges and improving the efficiency and effectiveness of monitoring systems.

3.2.5 Federated Learning

Federated Learning (FL) is increasingly being used to analyze distributed environmental data while preserving data privacy and security. This technique allows AI models to be trained across decentralized data sources, such as sensors or monitoring stations, without requiring data to be centrally stored. In the context of air quality monitoring, FL has shown significant promise. Xu et al. proposed a federated deep learning approach for air quality prediction using data from multiple monitoring stations [83]. Their study demonstrated that FL models could achieve comparable performance to centralized models while ensuring better data privacy and reducing data transmission requirements. This approach is particularly valuable for large-scale air quality monitoring networks where data sharing might be restricted due to privacy concerns or regulatory requirements. Federated learning (FL) has emerged as a promising approach for air pollution prediction in smart city applications, addressing challenges associated with centralized data processing. Nguyen and Zettsu proposed a spatially distributed FL framework utilizing Convolutional Recurrent Neural Networks (CRNN) for predicting oxidant warning levels across different spatial areas [84]. This approach allows for cooperative training among participants from various cities and prefectures without the need to centralize sensitive environmental data. The CRNN models are trained locally to capture spatial–temporal information specific to each area, while a global model aggregates knowledge from all participating regions. This methodology not only preserves data privacy but also reduces latency issues associated with transmitting large volumes of IoT sensor data to central servers. The study demonstrated that the common parts of the CRNN can be fused globally, while adaptive structures at the predictive part can accommodate different configurations of local environmental monitoring stations. Furthermore, the research showed that new participating areas could quickly obtain optimized local models by leveraging the transferred common global model, highlighting the scalability and efficiency of the FL approach in environmental monitoring applications. One of the key advantages of FL in environmental monitoring is its ability to handle privacy concerns while enabling collaborative learning. Alwabli (2024) presented an FL framework for predicting the Air Quality Index (AQI) using data from multiple Internet of Things (IoT) sensors deployed in urban areas [85]. This approach facilitates model training using diverse sensor data while maintaining data privacy at each source, with only model updates shared centrally. This is particularly valuable for smart city initiatives, where real-time, scalable, and efficient air quality monitoring systems are crucial.

3.2.6 Graph Neural Networks (GNNs)

Graph Neural Networks (GNNs) are indeed gaining significant traction in environmental monitoring due to their ability to effectively model and analyze complex spatial and topological structures inherent in environmental data. This advanced technique has shown promise in various environmental applications, particularly in areas where data can be naturally represented as graphs, such as sensor networks, river basins, or urban landscapes. In the realm of flood prediction, GNNs have demonstrated considerable potential. Kazadi et al. proposed FloodGNN, a graph neural network model for flood prediction that operates on both spatial and temporal dimensions [86]. Their approach, which processes water flow velocities as vector features, showed promising results in capturing complex flood patterns and outperformed traditional RNN-based models. GNNs have also been applied to air quality monitoring and prediction. Terroso-Saenz et al. introduced a novel GNN framework for nationwide air pollution forecasting that can handle heterogeneous settings with different types of monitoring stations [87]. Their approach captures similarities among stations related to land use and primary pollution sources, demonstrating the adaptability of GNNs to complex, large-scale environmental monitoring scenarios. In the context of soil quality assessment, particularly in predicting heavy metal concentrations, GNNs have shown promising results. Zha et al. developed a Multi-Scale Attention-based Graph Neural Network for Heavy Metal Prediction (MSA-GNN-HMP) model [88]. Tested in the Pearl River Basin, China, their model demonstrated superior performance in predicting soil heavy metal content compared to traditional methods, showcasing the potential of GNNs in addressing complex environmental challenges. In the context of water quality monitoring, Iwaniak et al. explored the use of GNNs for spatial analysis in urban areas [89]. Their study demonstrated how GNNs can effectively process graph representations of spatial data, offering new possibilities for analyzing urban environmental quality, including water resources. These applications demonstrate the versatility and effectiveness of GNNs in environmental monitoring. By leveraging the inherent spatial and topological structures in environmental data, GNNs offer a powerful tool for analyzing complex environmental interactions, predicting outcomes, and informing decision-making processes in environmental management and conservation efforts.

3.2.7 Anomaly Detection with AI

AI-powered anomaly detection is crucial for identifying unusual or unexpected events in environmental data. Techniques such as unsupervised learning and autoencoders are commonly employed to detect anomalies in time-series data from monitoring stations. These methods are particularly effective in early warning systems for detecting sudden spikes in pollution levels, temperature, or other environmental hazards. The ability to detect anomalies in real time can help authorities react more quickly to environmental threats. In the context of air quality monitoring, anomaly detection plays a vital role in improving data quality and subsequent analysis. Rollo et al. proposed AIrSense, a comprehensive framework that integrates anomaly detection, data cleaning, and repairing methodologies as pre-processing steps before applying calibration models to raw sensor readings [90]. Their approach combines multiple algorithms to enhance the robustness of anomaly detection, including a univariate sliding window method, a multivariate forgetting factor iterative data capture method, and an algorithm based on dependencies among pollutants and environmental measurements [90]. For river network systems, graph neural network-based approaches have shown promising results for anomaly detection. Buchhorn et al. applied the Graph Deviation Network (GDN) model to capture complex inter-sensor relationships and predict future sensor behaviour in river networks [91]. This approach uses sensor embeddings to learn a graph representation of the system and employs graph attention-based forecasting. Anomalies are identified when error scores exceed a calculated threshold, allowing for the detection of even small deviations that might be overlooked by traditional methods [91]. In indoor air quality monitoring, Long Short-Term Memory (LSTM) autoencoders have been employed for unsupervised outlier detection. Park et al. presented an LSTM-AE-based outlier detection framework that can detect various unpredictable abnormal situations by training deep neural networks on multi-feature data pools that change frequently depending on the indoor environment and timestamp [92]. Their approach combines reconstruction error-based detection with a latent feature class-assisted vector machine detector to create a robust ensemble model [92].

The integration of AI and graph neural networks has also advanced environmental monitoring capabilities in soil quality assessment. Zhang and Yang introduced a

Multi-Scale Attention-based Graph Neural Network for Heavy Metal Prediction (MSA-GNN-HMP) model [88]. Tested in the Pearl River Basin, China, their model demonstrated superior performance in predicting soil heavy metal content compared to traditional methods, showcasing the potential of GNNs in addressing complex environmental challenges [88]. These advancements in AI-powered anomaly detection represent significant progress in environmental monitoring, offering improved accuracy, efficiency, and timeliness in identifying and responding to environmental threats. As these technologies continue to evolve, they promise to enhance our ability to protect ecosystems, ensure public health, and manage environmental resources more effectively.

3.2.8 Explainable AI (XAI)

One of the challenges in environmental monitoring is ensuring that AI models are interpretable by policymakers and stakeholders. Explainable AI (XAI) techniques focus on making the decision-making processes of AI models transparent, allowing for greater trust and adoption. This is particularly important in regulatory contexts where decisions must be justified based on understandable parameters. Tasioulis and Karatzas highlight the crucial role of Explainable Artificial Intelligence (XAI) techniques, particularly SHAP (Shapley Additive Explanations) and LIME (Local Interpretable Model-Agnostic Explanations), in enhancing the transparency and interpretability of complex machine learning models used in air quality research [93]. These XAI methodologies have been effectively applied to model air pollutant concentrations worldwide, providing robust and comprehensible outcomes. In the context of environmental monitoring, SHAP and LIME help researchers and policymakers understand the factors influencing air quality predictions, identify important variables affecting pollutant concentrations, and explain specific air quality events or anomalies. By making complex models more interpretable, these techniques address the "black box" nature of advanced algorithms, enhancing the reliability and trustworthiness of air quality models. The authors emphasize that despite the transformative potential of SHAP, LIME, and other XAI techniques in improving the precision and transparency of environmental monitoring systems, a significant portion of air quality research still overlooks their implementation, potentially leading to biases and redundancies in datasets. They stress the need for continued integration and development of these XAI techniques to advance the field of environmental science and support more informed decision-making processes in air quality management [93]. Papadakis et al. discussed the opportunities and challenges of using data-driven, evidence-based decision-making in public policy through the analysis of large volumes of policy-related data from various sources such as e-services, mobile apps, and social media [94]. The authors highlight how machine learning (ML) and artificial intelligence (AI) technologies can facilitate the analysis of these large datasets, enabling policymakers to shift towards more data-driven decisions. The authors present a reference architecture and data mining framework for developing robust and unbiased systems for public policymaking. They also discuss the application of explainable AI techniques to make the decision-making processes of AI models more transparent and interpretable [94].

4 Application of AI in Various Fields of Environmental Monitoring

By leveraging machine learning algorithms, deep learning models, and reinforcement learning, researchers have successfully enhanced the accuracy and speed of environmental assessments, addressing key challenges such as pollution detection, climate change forecasting, and resource optimization. Table 1 summarizes historical research efforts that highlight the application of AI techniques across diverse environmental monitoring domains, showcasing the significant advancements made in this field.

5 Challenges Associated with Implementing AI in Environmental Monitoring

Technical Challenges

Implementing AI in environmental monitoring faces several technical hurdles. The complexity of environmental systems poses a significant challenge, as these systems are highly dynamic and involve numerous interacting variables, making it difficult for AI models to capture all relevant factors accurately [52]. Another major issue is model interpretability, particularly with deep learning models often described as "black boxes". This lack of transparency can hinder understanding of how these models arrive at their predictions, which is crucial for environmental decision-making [106]. Additionally, AI models processing large-scale environmental data require substantial computational resources, which can be a limiting factor for organizations or regions with limited access to high-performance computing infrastructure [79]. Figure 2 shows the key points on the challenges in implementing AI for environmental monitoring.

Data-Related Challenges

The quality and availability of data present significant challenges in AI implementation for environmental monitoring. AI models require large, high-quality datasets for training, but environmental data can often be sparse, inconsistent, or of varying quality, affecting model performance [107]. Data integration is another hurdle, as environmental monitoring involves diverse data types from various sources such as

Table 1 Summary of historical research work on the application of AI techniques in various environmental monitoring aspects

Case study/project name	Location	AI technique used	Environmental focus	Data sources	Key outcomes	Challenges addressed
Real-time air pollution health risk mapping [95]	Urban areas	Neural network models	Air pollution monitoring	Mobile and stationary air pollution sensors	Real-time health risk assessment	Integrating diverse data sources for comprehensive monitoring
Hydrological drought forecasting [96]	Gidra River, Slovakia	Artificial neural networks (ANNs), support vector machine (SVM)	Water resource management	Daily average discharge data	100% accuracy in predicting hydrological drought	Early classification of hydrological situations for irrigation planning and drought mitigation
Deep learning for hydrologic science [97]	Global	Deep learning	Hydrologic modeling and knowledge discovery	Various hydrologic data sources	Opening new avenues for knowledge discovery in hydrology	Integrating deep learning with traditional hydrologic sciences
Sustainable solid waste management [98]	Australia	Various AI techniques	Waste management	Municipal waste data, IoT sensors	Optimized waste collection and recycling	Improving waste management efficiency
Impact of transport modes on urban air quality [99]	Barcelona, Spain	Multilayer perceptron regressor	Air quality prediction and impact assessment	Road, maritime, and air traffic data; meteorological records	Accurate prediction of pollutant concentrations; Estimation of individual transport mode impacts; Evaluation of Low Emission Zone effectiveness	Overcoming limitations of classical pollutant dispersion models; Predicting air quality without detailed emission inventories

(continued)

Table 1 (continued)

Case study/project name	Location	AI technique used	Environmental focus	Data sources	Key outcomes	Challenges addressed
Emulating process-based water quality modeling [49]	Brusdalsvatnet lake, Ålesund, Norway	Long short-term memory (LSTM)	Water quality prediction in drinking water sources	Meteorological and hydrological measurements, process-based model results	Accurate prediction of temperature, *E. coli*, faecal coliforms, zinc, and lead concentrations; Spatial and temporal water quality predictions; R^2 ranges of 0.78–0.95 for temperature predictions	Simplifying complex hydrodynamic and water quality modeling; Enabling real-time simulations when integrated with field sensors
Climate model analysis [100]	Global	Machine learning feature analysis	Climate change modeling	Climate model outputs, observational data	Improved understanding of climate model discrepancies	Enhancing climate change predictions
AI-assisted air quality monitoring system for smart cities [50]	Selangor, Malaysia (Petaling Jaya, Banting, Klang, Shah Alam)	Ada Boost, SVR, RF, KNN, MLP regressor, LSTM (with feature optimization)	Air quality prediction	Air quality data (PM2.5, PM10, O_3, CO), meteorological data (wind speed, wind direction)	Best performance with feature-optimized LSTM (R^2 up to 0.998 for PM10 and PM2.5); Identification of key predictors (PM2.5, PM10, NO_2, humidity, wind speed)	Balancing urbanization and environmental sustainability; Providing data-driven solutions for smart city air quality management

(continued)

Table 1 (continued)

Case study/project name	Location	AI technique used	Environmental focus	Data sources	Key outcomes	Challenges addressed
Environmental justice and AI in urban air pollution monitoring [51]	Global perspective	Various AI and ML approaches	Urban air pollution monitoring and environmental justice	Air pollution data, public health surveys	Highlighting the potential and limitations of AI/ML in air quality monitoring; Emphasizing the need for balanced public and state control; Addressing environmental justice concerns	Balancing economic costs with monitoring needs; Ensuring reliability of inflexible monitoring systems; Avoiding concept substitution in smart city development
LSTM models for streamflow prediction in contrasting hydroclimate regions [101]	Great Basin and New England regions, USA	Long short-term memory (LSTM) networks	Daily streamflow prediction	CAMELS (catchment attributes and meteorology for large-sample studies) dataset	Regional models performed better in New England; Local models performed better in Great Basin; Three-layer LSTM architecture didn't significantly improve predictions	Assessing LSTM model applicability across different hydroclimate regions; Comparing local, regional, and grand LSTM models

(continued)

Table 1 (continued)

Case study/project name	Location	AI technique used	Environmental focus	Data sources	Key outcomes	Challenges addressed
Street-level traffic-related air pollutant prediction [102]	Hong Kong	Deep learning	Fine-scale traffic-related air pollution (TRAP) prediction	Mobile air quality sensors on buses, Google real-time traffic status, various urban features	Accurate prediction of NO and NO_2 concentrations (R^2 of 0.72 and 0.69 respectively); Improved model performance by 9–17% with traffic data; Insights on feature importance and interactions	Predicting fine-scale TRAP in complex urban environments; Incorporating real-time traffic data for improved accuracy
c-Air: mobile microscopy for air quality monitoring [103]	Los Angeles, USA	Machine learning with computational lens-free microscopy	Particulate Matter (PM) quantification and sizing	Microscopic images of aerosols in air	High-throughput PM quantification with 93% sizing accuracy; Correlation with EPA-approved devices; Mapping of air quality around LAX airport	Portable, cost-effective, and rapid air quality monitoring; Detection of specific particles like pollen and mold

(continued)

Table 1 (continued)

Case study/project name	Location	AI technique used	Environmental focus	Data sources	Key outcomes	Challenges addressed
Review of big data and ML approaches for air quality prediction [104]	Global perspective	Various (AI, decision trees, deep learning)	Air quality evaluation and prediction	IoT sensors, environmental sensing networks	Comprehensive review of big data and ML techniques for air quality forecasting; Insights on future research needs	Multi-dimensional factors affecting air quality; Real-time monitoring for smart cities
IoT-based air quality monitoring system [105]	India (with focus on New Delhi)	Machine learning (linear regression)	Air quality monitoring and analysis	MQ135 and MQ7 sensors, IoT platform (Thingspeak)	Real-time air quality monitoring; Public awareness through open dashboard; Cost-effective solution; Machine learning analysis of air quality trends	Urban air pollution monitoring; Public awareness; Cost reduction in air quality monitoring

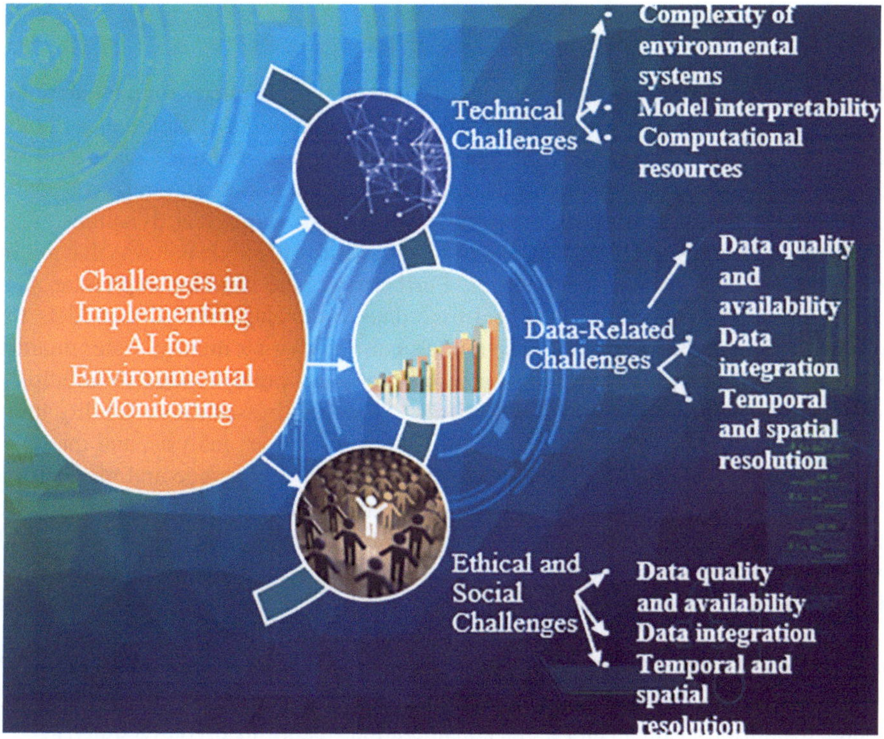

Fig. 2 Challenges in implementing AI for environmental monitoring

satellite imagery, sensor networks, and citizen science. Integrating these heterogeneous data sources can be challenging for AI systems [108]. Furthermore, environmental data often vary in temporal and spatial resolution, complicating the development of AI models that need to work across different scales [109].

Ethical and Social Challenges

The implementation of AI in environmental monitoring also raises several ethical and social concerns. Privacy is a significant issue, as environmental monitoring can involve collecting data that may infringe on individual privacy, especially in urban environments [110]. Equity and accessibility present another challenge, as AI implementation in environmental monitoring may not be equally accessible to all regions or communities, potentially exacerbating existing inequalities in environmental management and protection [111]. Trust and acceptance of AI-based environmental monitoring systems among stakeholders, including policymakers and the public, is crucial for effective implementation [112]. Lastly, the rapid development of AI technologies often outpaces regulatory frameworks, creating challenges in governing the use of AI in environmental monitoring and decision-making [110].

6 Conclusion

This chapter has explored the transformative potential of AI in environmental monitoring, highlighting its ability to overcome the limitations of traditional methods. The sources underscore how AI excels in processing vast, complex datasets from diverse sources, including satellite imagery, sensor networks, and citizen science initiatives, enabling a more nuanced understanding of ecological dynamics. This capability facilitates real-time analysis and identification of subtle patterns that might be missed by conventional approaches. Furthermore, the sources emphasize AI's capacity to analyze historical and real-time data to develop predictive models for various environmental factors. These models can forecast air quality, water quality, and species distribution, empowering researchers and policymakers to anticipate and mitigate environmental risks proactively. AI also addresses the scalability limitations of traditional methods by automating data collection, analysis, and interpretation processes. This enables more extensive monitoring coverage and facilitates a broader, more comprehensive understanding of environmental changes. However, the sources acknowledge the challenges in seamlessly integrating AI into environmental monitoring frameworks. Technical hurdles include the complexity of environmental systems, ensuring model interpretability (especially with "black box" deep learning models), and the need for substantial computational resources. Data-related challenges involve ensuring data quality, consistency, and availability for training AI models, as well as integrating heterogeneous data sources with varying temporal and spatial resolutions. Ethical and social considerations regarding privacy, equity, accessibility, and trust in AI-based decisions also require careful attention. Establishing robust regulatory frameworks that keep pace with rapid AI advancements is essential for responsible implementation. Hence, the sources presented in the chapter presents AI as a pivotal tool in revolutionizing environmental monitoring. While challenges remain, the potential benefits of AI in understanding, predicting, and mitigating environmental changes are undeniable. Future research and development should focus on addressing these challenges while harnessing AI's power to create a more sustainable and resilient future.

References

1. Roy, H. E., Pocock, M. J., Preston, C. D., Roy, D. B., Savage, J., Tweddle, J. C., & Robinson, L. D. (2012). *Understanding citizen science and environmental monitoring: Final report on behalf of UK Environmental Observation Framework.*
2. Markauskaite, L., Marrone, R., Poquet, O., Knight, S., Martinez-Maldonado, R., Howard, S., Tondeur, J., De Laat, M., Shum, S. B., & Gašević, D. (2022). Rethinking the entwinement between artificial intelligence and human learning: What capabilities do learners need for a world with AI? *Computers and Education: Artificial Intelligence, 1*(3), 100056.
3. Collins, H. (2021). The science of artificial intelligence and its critics. *Interdisciplinary Science Reviews, 46*(1–2), 53–70.

4. Viegas, C. V., Bond, A., Ribeiro, J. L. D., & Selig, P. M. (2013). A review of environmental monitoring and auditing in the context of risk: Unveiling the extent of a confused relationship. *Journal of Cleaner Production, 47*, 165–173.
5. Khankhoje, R. *AI in test automation: Overcoming challenges, embracing imperatives.*
6. Pettorelli, N., Laurance, W. F., O'Brien, T. G., Wegmann, M., Nagendra, H., & Turner, W. (2014). Satellite remote sensing for applied ecologists: Opportunities and challenges. *Journal of Applied Ecology, 51*(4), 839–848.
7. Koh, L. P., & Wich, S. A. (2012). Dawn of drone ecology: Low-cost autonomous aerial vehicles for conservation. *Tropical Conservation Science, 5*(2), 121–132.
8. Zhang, L., & Zhang, L. (2022). Artificial intelligence for remote sensing data analysis: A review of challenges and opportunities. *IEEE Geoscience and Remote Sensing Magazine, 10*(2), 270–294.
9. Hampton, S. E., Strasser, C. A., Tewksbury, J. J., Gram, W. K., Budden, A. E., Batcheller, A. L., Duke, C. S., & Porter, J. H. (2013). Big data and the future of ecology. *Frontiers in Ecology and the Environment, 11*(3), 156–162.
10. Willcock, S., Martínez-López, J., Hooftman, D. A., Bagstad, K. J., Balbi, S., Marzo, A., Prato, C., Sciandrello, S., Signorello, G., & Voigt, B. (2018). Machine learning for ecosystem services. *Ecosystem Services, 33*, 165–174.
11. Christin, S., Hervet, É., & Lecomte, N. (2019). Applications for deep learning in ecology. *Methods in Ecology and Evolution, 10*(10), 1632–1644.
12. Tabak, M. A., Norouzzadeh, M. S., Wolfson, D. W., Sweeney, S. J., VerCauteren, K. C., Snow, N. P., Halseth, J. M., Di Salvo, P. A., Lewis, J. S., & White, M. D. (2019). Machine learning to classify animal species in camera trap images: Applications in ecology. *Methods in Ecology and Evolution, 10*(4), 585–590.
13. Himeur, Y., Rimal, B., Tiwary, A., & Amira, A. (2022). Using artificial intelligence and data fusion for environmental monitoring: A review and future perspectives. *Information Fusion, 86*, 44–75.
14. Akter, M. S. (2024). Harnessing technology for environmental sustainability: Utilizing AI to tackle global ecological challenge. *Journal of Artificial Intelligence General Science (JAIGS), 2*(1), 61–70. ISSN: 3006-4023.
15. Shabani, N. B. (2023). *A comparative evaluation of AI-based methods and traditional approaches for air quality monitoring: Analyzing pros and cons.*
16. Krtolica, I., & Medojević, M. (2024). Enhancing precision in artificial intelligence–based water quality prediction: The advantages of hybrid modeling approaches—Review. In *2024 9th International Conference on Smart and Sustainable Technologies (SpliTech)*. IEEE.
17. Amiri, Z., Heidari, A., & Navimipour, N. J. (2024). Comprehensive survey of artificial intelligence techniques and strategies for climate change mitigation. *Energy*, 132827.
18. Zong, Z., & Guan, Y. (2024). AI-driven intelligent data analytics and predictive analysis in Industry 4.0: Transforming knowledge, innovation, and efficiency. *Journal of the Knowledge Economy*, 1–40.
19. Pattyam, S. P. (2021). AI-driven data science for environmental monitoring: Techniques for data collection, analysis, and predictive modeling. *Australian Journal of Machine Learning Research & Applications, 1*(1), 132–169.
20. Lee, S.-H., & Lee, M.-J. (2021). A study of establishment and application algorithm of artificial intelligence training data on land use/cover using aerial photograph and satellite images. *Korean Journal of Remote Sensing, 37*(5), 871–884.
21. Singh, D., Dahiya, M., Kumar, R., & Nanda, C. (2021). Sensors and systems for air quality assessment monitoring and management: A review. *Journal of Environmental Management, 289*, 112510.
22. AlShehhi, A., & Welsch, R. (2023). Artificial intelligence for improving Nitrogen Dioxide forecasting of Abu Dhabi environment agency ground-based stations. *Journal of Big Data, 10*(1), 92.
23. Xu, K. (2023). Artificial intelligence unlocks ecological environment governance—Smart statistical monitoring based on meteorology. *Multimedia Tools and Applications, 82*(14), 21613–21625.

24. Lotrecchiano, N., Barletta, D., Poletto, M., & Sofia, D. (2022). Artificial intelligence for the pollution source identification. *Chemical Engineering Transactions, 96*, 439–444.
25. Masood, A., & Ahmad, K. (2021). A review on emerging artificial intelligence (AI) techniques for air pollution forecasting: Fundamentals, application and performance. *Journal of Cleaner Production, 322*, 129072.
26. Uddin, M. G., Nash, S., Diganta, M. T. M., Rahman, A., & Olbert, A. I. (2022). Robust machine learning algorithms for predicting coastal water quality index. *Journal of Environmental Management, 321*, 115923.
27. John, K., Abraham Isong, I., Michael Kebonye, N., Okon Ayito, E., Chapman Agyeman, P., & Marcus Afu, S. (2020). Using machine learning algorithms to estimate soil organic carbon variability with environmental variables and soil nutrient indicators in an alluvial soil. *Land, 9*(12), 487.
28. Zhao, J., & Wan, S. (2023). Artificial intelligence and hyperspectral modeling for soil management. *Innovation for environmentally-friendly food production and food safety in China* (pp. 67–91). Springer.
29. Lehikoinen, A., Olsson, J., Bergström, L., Bergström, U., Bryhn, A., Fredriksson, R., & Uusitalo, L. (2019). Evaluating complex relationships between ecological indicators and environmental factors in the Baltic Sea: A machine learning approach. *Ecological Indicators, 101*, 117–125.
30. Pandey, D. K., Hunjra, A. I., Bhaskar, R., & Al-Faryan, M. A. S. (2023). Artificial intelligence, machine learning and big data in natural resources management: A comprehensive bibliometric review of literature spanning 1975–2022. *Resources Policy, 86*, 104250.
31. Shivaprakash, K. N., Swami, N., Mysorekar, S., Arora, R., Gangadharan, A., Vohra, K., Jadeyegowda, M., & Kiesecker, J. M. (2022). Potential for artificial intelligence (AI) and machine learning (ML) applications in biodiversity conservation, managing forests, and related services in India. *Sustainability, 14*(12), 7154.
32. Chaturvedi, V., & de Vries, W. T. (2021). Machine learning algorithms for urban land use planning: A review. *Urban Science, 5*(3), 68.
33. Antonelli, A., Dhanjal-Adams, K. L., & Silvestro, D. (2023). Integrating machine learning, remote sensing and citizen science to create an early warning system for biodiversity. *Plants, People, Planet, 5*(3), 307–316.
34. Biber, E. (2013). The challenge of collecting and using environmental monitoring data. *Ecology and Society, 18*(4).
35. Marvin, D. C., Koh, L. P., Lynam, A. J., Wich, S., Davies, A. B., Krishnamurthy, R., Stokes, E., Starkey, R., & Asner, G. P. (2016). Integrating technologies for scalable ecology and conservation. *Global Ecology and Conservation, 7*, 262–275.
36. Kumar, A., Kim, H., & Hancke, G. P. (2012). Environmental monitoring systems: A review. *IEEE Sensors Journal, 13*(4), 1329–1339.
37. Rodriguez-Gil, J. L., Sauto, J. S. S., González-Alonso, S., Sanchez, P. S., Valcarcel, Y., & Catalá, M. (2013). Development of cost-effective strategies for environmental monitoring of irrigated areas in Mediterranean regions: Traditional and new approaches in a changing world. *Agriculture, Ecosystems & Environment, 181*, 41–49.
38. Gordon, S. I. (2012). *Computer models in environmental planning*. Springer Science & Business Media.
39. Gregory, R., Failing, L., Harstone, M., Long, G., McDaniels, T., & Ohlson, D. (2012). *Structured decision making: A practical guide to environmental management choices*. Wiley.
40. Fascista, A. (2022). Toward integrated large-scale environmental monitoring using WSN/UAV/Crowdsensing: A review of applications, signal processing, and future perspectives. *Sensors, 22*(5), 1824.
41. Tandon, N., & Shaheen, M. (2024). AI-driven innovations for environmental sustainability across sectors. In *Maintaining a sustainable world in the nexus of environmental science and AI* (pp. 429–454). IGI Global.
42. Popescu, S. M., Mansoor, S., Wani, O. A., Kumar, S. S., Sharma, V., Sharma, A., Arya, V. M., Kirkham, M., Hou, D., & Bolan, N. (2024). Artificial intelligence and IoT driven technologies

for environmental pollution monitoring and management. *Frontiers in Environmental Science, 12*, 1336088.
43. Gibert, K., Izquierdo, J., Sànchez-Marrè, M., Hamilton, S. H., Rodríguez-Roda, I., & Holmes, G. (2018). Which method to use? An assessment of data mining methods in Environmental Data Science. *Environmental Modelling & Software, 110*, 3–27.
44. Konya, A., & Nematzadeh, P. (2024). Recent applications of AI to environmental disciplines: A review. *Science of The Total Environment, 906*, 167705.
45. Sani, S. A. (2023). Drawbacks of traditional environmental monitoring systems. *TMP Universal Journal of Research and Review Archives, 2*(2).
46. Holloway, J., Mengersen, K., & Helmstedt, K. (2018). Spatial and machine learning methods of satellite imagery analysis for sustainable development goals. In *Proceedings of the 16th Conference of International Association for Official Statistics (IAOS)*. International Association for Official Statistics (IAOS).
47. Hino, M., Benami, E., & Brooks, N. (2018). Machine learning for environmental monitoring. *Nature Sustainability, 1*(10), 583–588.
48. Kulikova, E., Sulimin, V., & Shvedov, V. (2023). Artificial intelligence for ambient air quality control. In *E3S Web of Conferences*. EDP Sciences.
49. Mohammed, H., Tornyeviadzi, H. M., & Seidu, R. (2022). Emulating process-based water quality modelling in water source reservoirs using machine learning. *Journal of Hydrology, 609*, 127675.
50. Neo, E. X., Hasikin, K., Lai, K. W., Mokhtar, M. I., Azizan, M. M., Hizaddin, H. F., & Razak, S. A. (2023). Artificial intelligence-assisted air quality monitoring for smart city management. *Peer J Computer Science, 9*, e1306.
51. Krupnova, T. G., Rakova, O. V., Bondarenko, K. A., & Tretyakova, V. D. (2022). Environmental justice and the use of artificial intelligence in urban air pollution monitoring. *Big Data and Cognitive Computing, 6*(3), 75.
52. Reichstein, M., Camps-Valls, G., Stevens, B., Jung, M., Denzler, J., Carvalhais, N., & Prabhat, F. (2019). Deep learning and process understanding for data-driven Earth system science. *Nature, 566*(7743), 195–204.
53. Tuia, D., Kellenberger, B., Beery, S., Costelloe, B. R., Zuffi, S., Risse, B., Mathis, A., Mathis, M. W., Van Langevelde, F., & Burghardt, T. (2022). Perspectives in machine learning for wildlife conservation. *Nature Communications, 13*(1), 1–15.
54. Lary, D. J., Alavi, A. H., Gandomi, A. H., & Walker, A. L. (2016). Machine learning in geosciences and remote sensing. *Geoscience Frontiers, 7*(1), 3–10.
55. Holloway, J., & Mengersen, K. (2018). Statistical machine learning methods and remote sensing for sustainable development goals: A review. *Remote Sensing, 10*(9), 1365.
56. Gomes, V. C., Queiroz, G. R., & Ferreira, K. R. (2020). An overview of platforms for big earth observation data management and analysis. *Remote Sensing, 12*(8), 1253.
57. Joppa, L. N. (2017). The case for technology investments in the environment. *Nature, 552*(7685), 325–328.
58. Faghmous, J. H., & Kumar, V. (2014). A big data guide to understanding climate change: The case for theory-guided data science. *Big Data, 2*(3), 155–163.
59. Stowell, D., Wood, M. D., Pamuła, H., Stylianou, Y., & Glotin, H. (2019). Automatic acoustic detection of birds through deep learning: The first bird audio detection challenge. *Methods in Ecology and Evolution, 10*(3), 368–380.
60. Zheng, Y., Liu, F., & Hsieh, H.-P. (2013). U-air: When urban air quality inference meets big data. In *Proceedings of the 19th ACM SIGKDD International Conference on Knowledge Discovery and Data Mining*.
61. Bonney, R., Cooper, C. B., Dickinson, J., Kelling, S., Phillips, T., Rosenberg, K. V., & Shirk, J. (2009). Citizen science: A developing tool for expanding science knowledge and scientific literacy. *BioScience, 59*(11), 977–984.
62. Zhang, X., Shu, K., Rajkumar, S. A., & Sivakumar, V. (2021). Research on deep integration of application of artificial intelligence in environmental monitoring system and real economy. *Environmental Impact Assessment Review, 86*, 106499.

63. Fang, S., Da Xu, L., Zhu, Y., Ahati, J., Pei, H., Yan, J., & Liu, Z. (2014). An integrated system for regional environmental monitoring and management based on internet of things. *IEEE Transactions on Industrial Informatics, 10*(2), 1596–1605.
64. Yao, Y. (2022). A review of the comprehensive application of big data, artificial intelligence, and internet of things technologies in smart cities. J. Comput. Method Eng. Appl. 1–10.
65. Manjakkal, L., Mitra, S., Petillot, Y. R., Shutler, J., Scott, E. M., Willander, M., & Dahiya, R. (2021). Connected sensors, innovative sensor deployment, and intelligent data analysis for online water quality monitoring. *IEEE Internet of Things Journal, 8*(18), 13805–13824.
66. Zare, A., Ablakimova, N., Kaliyev, A. A., Mussin, N. M., Tanideh, N., Rahmanifar, F., & Tamadon, A. (2024). An update for various applications of Artificial Intelligence (AI) for detection and identification of marine environmental pollutions: A bibliometric analysis and systematic review. *Marine Pollution Bulletin, 206*, 116751.
67. Raihan, A. (2023). Artificial intelligence and machine learning applications in forest management and biodiversity conservation. *Natural Resources Conservation and Research, 6*(2), 3825.
68. Collins, A. C. (2024). Harnessing innovations in AI and robotics for environmental conservation: A comprehensive overview.
69. Tomar, P., & Melkania, N. Environmental sustainability through artificial intelligence: A review.
70. Zhao, P., & Zettsu, K. (2018). Convolution recurrent neural networks for short-term prediction of atmospheric sensing data. In *2018 IEEE International Conference on Internet of Things (iThings) and IEEE Green Computing and Communications (GreenCom) and IEEE Cyber, Physical and Social Computing (CPSCom) and IEEE Smart Data (SmartData)*. IEEE.
71. Yang, S. (2023). *Attention-based deep neural networks for the forecasting and reconstruction of an environmental image series*. University of British Columbia.
72. Yang, Y., Cui, Q., Cheng, R., Huo, A., & Wang, Y. (2023). Retrieval of soil heavy metal content for environment monitoring in mining area via transfer learning. *Sustainability, 15*(15), 11765.
73. Méndez-Zambrano, P. V., Tierra Pérez, L. P., Ureta Valdez, R. E., & Flores Orozco, Á. P. (2023). Technological innovations for agricultural production from an environmental perspective: A review. *Sustainability, 15*(22), 16100.
74. Fijani, E., Barzegar, R., Deo, R., Tziritis, E., & Skordas, K. (2019). Design and implementation of a hybrid model based on two-layer decomposition method coupled with extreme learning machines to support real-time environmental monitoring of water quality parameters. *Science of the Total Environment, 648*, 839–853.
75. Wen, C., Liu, S., Yao, X., Peng, L., Li, X., Hu, Y., & Chi, T. (2019). A novel spatiotemporal convolutional long short-term neural network for air pollution prediction. *Science of the Total Environment, 654*, 1091–1099.
76. Wen CongCong, W. C., Liu ShuFu, L. S., Yao XiaoJing, Y. X., Peng Ling, P. L., Li Xiang, L. X., Hu Yuan, H. Y., & Chi TianHe, C. T. (2019). *A novel spatiotemporal convolutional long short-term neural network for air pollution prediction.*
77. Zhang, D., Lindholm, G., & Ratnaweera, H. (2018). Use long short-term memory to enhance Internet of Things for combined sewer overflow monitoring. *Journal of Hydrology, 556*, 409–418.
78. Liakos, K. G., Busato, P., Moshou, D., Pearson, S., & Bochtis, D. (2018). Machine learning in agriculture: A review. *Sensors, 18*(8), 2674.
79. Rolnick, D., Donti, P. L., Kaack, L. H., Kochanski, K., Lacoste, A., Sankaran, K., Ross, A. S., Milojevic-Dupont, N., Jaques, N., & Waldman-Brown, A. (2022). Tackling climate change with machine learning. *ACM Computing Surveys (CSUR), 55*(2), 1–96.
80. Castelletti, A., Galelli, S., Restelli, M., & Soncini-Sessa, R. (2010). Tree-based reinforcement learning for optimal water reservoir operation. Water Resources Research, 46(9).
81. Ayus, I., Natarajan, N., & Gupta, D. (2023). Comparison of machine learning and deep learning techniques for the prediction of air pollution: A case study from China. *Asian Journal of Atmospheric Environment, 17*(1), 4.

82. Kober, J., Bagnell, J. A., & Peters, J. (2013). Reinforcement learning in robotics: A survey. *The International Journal of Robotics Research, 32*(11), 1238–1274.
83. Xu, D., Li, T., Li, Y., Su, X., Tarkoma, S., Jiang, T., Crowcroft, J., & Hui, P. (2020). *Edge intelligence: Architectures, challenges, and applications*. arXiv preprint arXiv:2003.12172.
84. Nguyen, D.-V., & Zettsu, K. (2021). Spatially-distributed federated learning of convolutional recurrent neural networks for air pollution prediction. In *2021 IEEE International Conference on Big Data (Big Data)*. IEEE.
85. Alwabli, A. (2024). Federated learning for privacy-preserving air quality forecasting using IoT sensors. *Engineering, Technology & Applied Science Research, 14*(4), 16069–16076.
86. Kazadi, A. N., Doss-Gollin, J., Sebastian, A., & Silva, A. (2022). *Flood prediction with graph neural networks*. Climate Change AI.
87. Terroso-Saenz, F., Morales-García, J., & Muñoz, A. (2024). Nationwide air pollution forecasting with heterogeneous graph neural networks. *ACM Transactions on Intelligent Systems and Technology, 15*(1), 1–19.
88. Zha, Y., & Yang, Y. (2024). Innovative graph neural network approach for predicting soil heavy metal pollution in the Pearl River Basin, China. *Scientific Reports, 14*(1), 16505.
89. Iwaniak, A., Maziarz, M., & Kaczmarek, I. (2023). Spatial analysis in GeoAI with the use of graph neural networks: Quality of life prediction in urban areas case. *Abstracts of the ICA, 6*, 102.
90. Rollo, F., Bachechi, C., & Po, L. (2023). Anomaly detection and repairing for improving air quality monitoring. *Sensors, 23*(2), 640.
91. Buchhorn, K., Santos-Fernandez, E., Mengersen, K., & Salomone, R. (2023). Graph neural network-based anomaly detection for river network systems. *F1000Research, 12*.
92. Park, J., Seo, Y., & Cho, J. (2023). Unsupervised outlier detection for time-series data of indoor air quality using LSTM autoencoder with ensemble method. *Journal of Big Data, 10*(1), 66.
93. Tasioulis, T., & Karatzas, K. (2023). Reviewing explainable artificial intelligence towards better air quality modelling. *Environmental informatics* (pp. 3–19). Springer.
94. Papadakis, T., Christou, I. T., Ipektsidis, C., Soldatos, J., & Amicone, A. (2024). Explainable and transparent artificial intelligence for public policymaking. *Data & Policy, 6*, e10.
95. Jerrett, M., Donaire-Gonzalez, D., Popoola, O., Jones, R., Cohen, R. C., Almanza, E., De Nazelle, A., Mead, I., Carrasco-Turigas, G., & Cole-Hunter, T. (2017). Validating novel air pollution sensors to improve exposure estimates for epidemiological analyses and citizen science. *Environmental Research, 158*, 286–294.
96. Almikaeel, W., Čubanová, L., & Šoltész, A. (2022). Hydrological drought forecasting using machine learning—Gidra river case study. *Water, 14*(3), 387.
97. Shen, C., Laloy, E., Elshorbagy, A., Albert, A., Bales, J., Chang, F.-J., Ganguly, S., Hsu, K.-L., Kifer, D., Fang, Z., & Opinions, H. E. S. S. (2018). Incubating deep-learning-powered hydrologic science advances as a community. *Hydrology and Earth System Sciences, 22*(11), 5639–5656.
98. Esmaeilian, B., Wang, B., Lewis, K., Duarte, F., Ratti, C., & Behdad, S. (2018). The future of waste management in smart and sustainable cities: A review and concept paper. *Waste Management, 81*, 177–195.
99. Fabregat, A., Vernet, A., Vernet, M., Vázquez, L., & Ferré, J. A. (2022). Using Machine Learning to estimate the impact of different modes of transport and traffic restriction strategies on urban air quality. *Urban Climate, 45*, 101284.
100. Hoffman, F. M., Hargrove, W. W., Jr., Erickson, D. J., III., & Oglesby, R. J. (2005). Using clustered climate regimes to analyze and compare predictions from fully coupled general circulation models. *Earth Interactions, 9*(10), 1–27.
101. Khand, K., & Senay, G. B. (2024). Evaluation of streamflow predictions from LSTM models in water-and energy-limited regions in the United States. *Machine Learning with Applications, 16*, 100551.
102. Wei, P., Hao, S., Shi, Y., Anand, A., Wang, Y., Chu, M., & Ning, Z. (2024). Combining Google traffic map with deep learning model to predict street-level traffic-related air pollutants in a complex urban environment. *Environment International*, 108992.

103. Wu, Y.-C., Shiledar, A., Li, Y.-C., Wong, J., Feng, S., Chen, X., Chen, C., Jin, K., Janamian, S., & Yang, Z. (2017). Air quality monitoring using mobile microscopy and machine learning. *Light: Science & Applications, 6*(9), e17046–e17046.
104. Kang, G. K., Gao, J. Z., Chiao, S., Lu, S., & Xie, G. (2018). Air quality prediction: Big data and machine learning approaches. *International Journal of Environmental Science and Development, 9*(1), 8–16.
105. Sai, K. B. K., Subbareddy, S. R., & Luhach, A. K. (2019). IOT based air quality monitoring system using MQ135 and MQ7 with machine learning analysis. *Scalable Computing: Practice and Experience, 20*(4), 599–606.
106. Roscher, R., Bohn, B., Duarte, M. F., & Garcke, J. (2020). Explainable machine learning for scientific insights and discoveries. *IEEE Access, 8*, 42200–42216.
107. Huntingford, C., Jeffers, E. S., Bonsall, M. B., Christensen, H. M., Lees, T., & Yang, H. (2019). Machine learning and artificial intelligence to aid climate change research and preparedness. *Environmental Research Letters, 14*(12), 124007.
108. Karpatne, A., Ebert-Uphoff, I., Ravela, S., Babaie, H. A., & Kumar, V. (2018). Machine learning for the geosciences: Challenges and opportunities. *IEEE Transactions on Knowledge and Data Engineering, 31*(8), 1544–1554.
109. Vicens-Miquel, M., Williams, D. D., & Tissot, P. E. (2024). Analysis of sandy beach morphology changes and inundation events from a high spatiotemporal resolution dataset. *Journal of Coastal Research.*
110. Vinuesa, R., Azizpour, H., Leite, I., Balaam, M., Dignum, V., Domisch, S., Felländer, A., Langhans, S. D., Tegmark, M., & Fuso Nerini, F. (2020). The role of artificial intelligence in achieving the sustainable development goals. *Nature Communications, 11*(1), 1–10.
111. Yigitcanlar, T., Mehmood, R., & Corchado, J. M. (2021). Green artificial intelligence: Towards an efficient, sustainable and equitable technology for smart cities and futures. *Sustainability, 13*(16), 8952.
112. Rane, N., Choudhary, S. P., & Rane, J. (2024). Acceptance of artificial intelligence: Key factors, challenges, and implementation strategies. *Journal of Applied Artificial Intelligence, 5*(2), 50–70.

Chapter 3
Optimizing Renewable Energy with AI: Environmental Monitoring and Predictive Maintenance Strategies

Akshay Garg, Abu Zar, Balendra V. S. Chauhan, and Siddharth Jain

Abstract Conventional energy sources play a significant role in meeting the energy demand globally, which is responsible for the economic growth worldwide. Consumption of fossil fuels results in a significant amount of greenhouse gases emissions and environmental impacts. In the last two decades, Renewable energy systems have attracted various researchers, scientists, and government policymakers for their sustainability potential. It is apparent that renewable energy systems assist in the mitigation of emissions into the environment, however, their significant negative impacts on environmental health cannot be ignored and are to be considered. The effects include pollution, greenhouse gases emissions, ecological disturbances, human health, ozone depletion, flooding, and natural resources depletion. Although these effects are caused much less harshly as compared to fossil fuels, nevertheless appropriate precautions are to be taken. Technology development has given rise to fields notably as Machine Learning (ML) and Artificial intelligence (AI). They do possess the power to change the renewable energy sector. Power firms may improve their projections, administer existing grid, and plan upgrades by utilizing AI. Renewables are without a doubt the energy of the future, yet one of its biggest problems is that it is unpredictable. AI has helped a lot in whether prediction, this forecast information is used by the energy providers to manage the power systems. The businesses generate and preserve sustainable power if the prognosis is favorable.

B. V. S. Chauhan
Centre for Earth Observation Science, University of Brighton, Brighton, United Kingdom
e-mail: b.chauhan@brighton.ac.uk; balendrachauhan8801@gmail.com

S. Jain
School of Mechnical Engineering, UPES, Dehradun, India
e-mail: siddharth.jain@ddn.upes.ac.in

A. Garg (✉)
Industrial Ecology, Chalmers University of Technology, Gothenburg, Sweden
e-mail: akgarg.2411@gmail.com

A. Zar
Renewable Energy Engineering, School of Water, Environment and Energy, Cranfield University, Cranfield, United Kingdom
e-mail: mech.abuzar@gmail.com

A. Vedrtnam et al., *Prospects of Artificial Intelligence in the Environment*,
https://doi.org/10.1007/978-981-96-6863-2_3

Power firms adjust overall load management depending on negative forecasts also. To maintain a continuous energy, companies prepare ahead of potential issue and turn to coal and oil for assistance. This chapter provides an overview of the social and environmental impacts of various renewable energy systems: solar, wind, hydro, and biomass energy. Implications of AI are reviewed and discussed with a curiosity to accomplish thirst of knowledge for how AI succors' renewable energy systems. Also, a strength, weakness, opportunity, and threat (SWOT) analysis is formulated for all the energy systems to assist the energy advisors in considering both positive and negative impacts of particular energy systems.

Keywords Artificial Intelligence · Renewable energy · Machine learning · Environment · SWOT

1 Introduction

The industrial evolution and economic growth of the world critically depends upon the reliable and affordable sources of energy [1]. But increasing the consumption of conventional sources of energy such as coal, oil and natural gas has increased the concern of exponential increase in the greenhouse effect. This has led the world to face major climate change impacts like high temperature, drought, floods, and uncertain weather, etc. [2]. To reduce carbon emissions, it has become necessary to reduce the use of fossil fuels and develop new technologies that are environmentally friendly and can play a major role in sustainable development. The establishment of the Paris Agreement has imposed a responsibility upon the nations to work on the mitigation of climate change by reducing greenhouse gases (GHGs) emissions [3]. Despite high consumption of fossil fuels, around 733 million people lack access to electricity in 2020 globally. People are obligate to use wood, biomass and kerosene oil for the purpose of cooking and heating. The pace of progress in electrifying the world has been slower in the last decade due to increase in population and difficulty in connecting the grid-in rural areas [4]. United nations have set 17 Sustainable Development Goals to be achieved by 2030, so as to protect the environment and ensure the welfare of the society and ecosystem. According to Sustainable Development Goals (SDG 7) set by the United Nations, electricity is to be provided globally to every individual by 2030 [5].

Due to growing demand of energy requirement, it has become essential to increase the energy generation sources. Increasing the consumption of fossil fuels to meet the energy demand would adversely affect the environment. Thus, to tackle the above crisis, renewable energy systems (RES) have been found as promising sources of energy due to no carbon emissions and environmentally friendly [6]. The world is gradually moving towards RES to meet the energy demand. It is clear from Fig. 1. that the share of fossil fuels in energy production is expected to decrease to 60% by 2050.

The renewable energy has been significantly developed in last two decades. There are majorly four RES that share a major part of meeting energy demand through renewable sources. These are solar, wind, hydro and biomass. It can clearly be observed from Fig. 2. that hydropower has the highest percentage of share in the energy contribution among RES. Although RES are better than fossil fuels but can be proven harmful for the environment if not managed properly [8]. Numerous studies have been conducted by researchers globally, to understand the lifecycle impact of different energy systems. It has been reported that different RES have their impact on environment and ecological systems during the whole lifecycle [9]. In this chapter all four RES are incorporated and discussed briefly about their potential and negative impacts. The chapter also elaborates the strength, weakness, opportunities and threat (SWOT) analysis of all the four energy systems.

Numerous studies are being done to improve the efficiency of RES and mitigating the environmental effects associated with them. Improvement in designs, technical advancements for increasing efficiency, integration of different systems to install hybrid systems and distribution systems are being studies thoroughly to establish the RES efficiently [11–14].

Recently, AI has gathered the attention of scientists as a powerful tool to enhance the energy systems performance and mitigate the challenges of GHGs. AI is a data-intensive technology that can be used to analyze the present data and predict the future data to check the feasibility of the new proposed systems. AI mainly works on

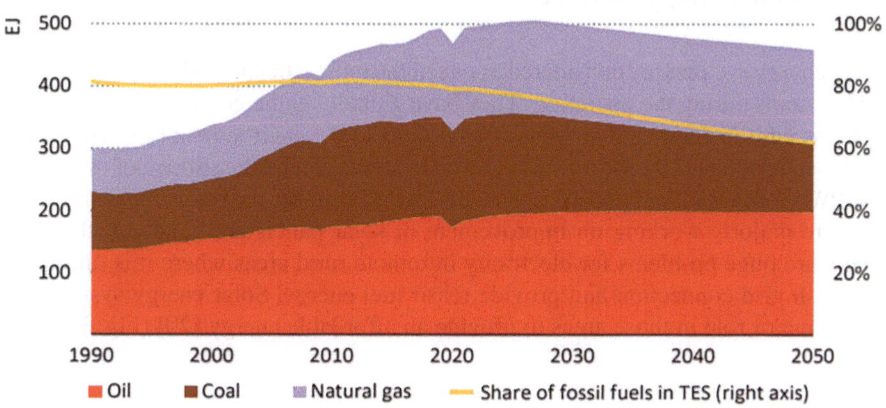

Fig. 1 Fossil fuel energy share by 2050 [7]

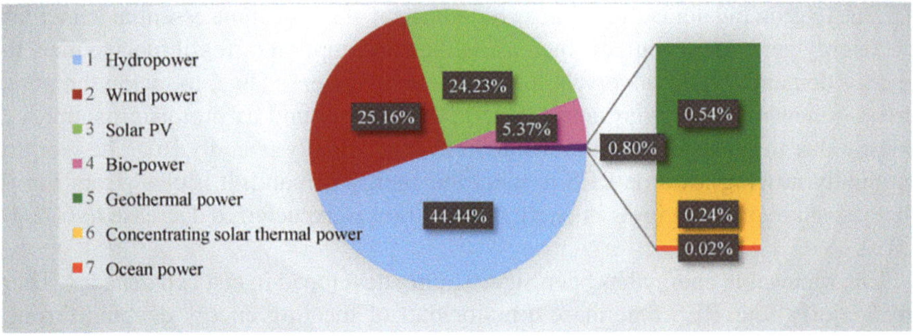

Fig. 2 Shares of renewable energy systems [10]

optimizing the system performance and has its applications in countless industries like health, food, energy, education and agriculture, etc. [15, 16]. This data can play a major role in calculating the uncertainties and projecting the future risks of energy demand and consumption. It is difficult to totally rely on the RES due to inconsistency in supply and demand conditions. This can be solved by training the data models and predicting the weather predictions of the location, the plant is installed especially in case of solar and wind energy [17]. AI would lead the energy sector to obtain an optimum combination of components both economically and technically and produce reliable systems that can reduce the investment cost in energy. Thus, this chapter reviews the role of different AI optimization techniques to enhance the RES power systems.

2 Renewable Energy Systems

2.1 Solar Energy Systems

Solar energy systems are considered as one of the most efficient and zero carbon emitting systems during the operation. They have globally gathered the attention as solar systems emit insignificant amount of emissions to the environment [18]. According to Hosenuzzaman [19], the power received from the sun by the surface of the earth is 140PW, from which only 36PW can be utilized. Currently, the researchers and scientists are majorly working on improvement of solar panels and related technology. There are huge problems for electricity in remote rural areas where it is difficult to establish grid connection and provide fossil fuel energy. Solar energy systems can play a major role in those areas to provide an affordable energy [20] (Fig. 3).

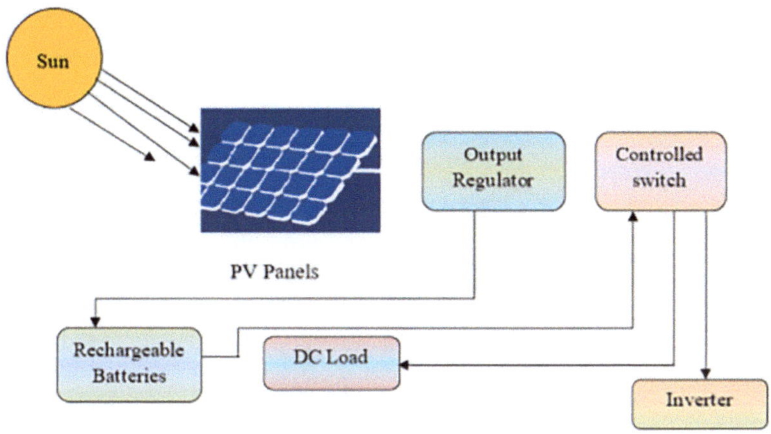

Fig. 3 Schematic diagram of a solar PV system [21]

2.2 *Environmental Impacts of Solar Energy*

Though net-zero emissions during the operation, there are concerns related to production, installation, and the disposal of the solar waste [22]. Solar systems are generally located on the land, and small countries with lack of land are not suitable for solar plant installation. If solar plants are located on agricultural land, it would affect the soil fertility and the cultivation capacity of the land. Thus, generating energy on the cost of food scarcity is not feasible [23]. A large amount of water is consumed by solar power plants to generate electricity, thus creating a high environmental impact. Solar power plants have been found to be a danger for water-energy-food nexus [24]. Photovoltaic cells contain numerous ingredients that are to be managed properly to cause less harm to the environment and human health. For example, solar cells usually contain toxic elements like tin, lead, mercury, and chromium that pollute the environment. Silicon is one the main ingredient of photovoltaic modules that can have adverse effect on human lungs if inhaled like dust [25]. Solar panels have a long life of 25 years, so the management of solar waste after end-of-life is going to be another environmental crisis for the world in near future [21].

2.3 *Wind Energy Systems*

Wind energy has been recognized as one of the most mature and eco-friendly energy sources after hydropower. The techno-economic analysis studied for the wind energy systems have proved the potential feasibility of the technology. It is one of the oldest sources of energy and can be installed both on terrestrial and aquatic environment [26]. The basic principle of working of a wind turbine works on converting the mechanical energy into electrical energy. The wind in the atmosphere rotates the

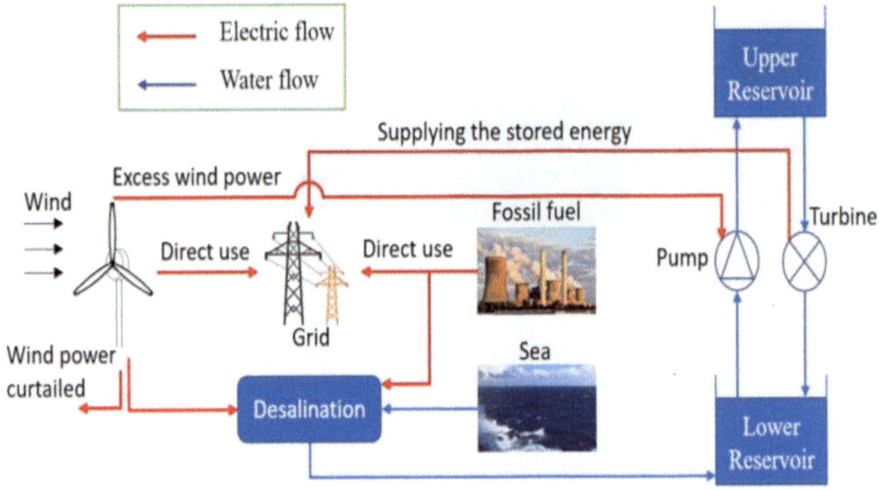

Fig. 4 Diagram of a wind energy system [29]

blades of a wind turbine. The turbine is coupled with a generator and the rotatory movement of the turbine generates electricity [27]. According to IRENA, wind energy has gradually gathered the attention of scientists after solar energy as the most promising renewable energy source as they do not produce any gaseous emissions such as CO_2, NOx, PM or soot during the operation [28]. Besides ease of operation and installation, some studies have reported numerous negative impacts on the environment (Fig. 4).

2.3.1 Environmental Impacts of Wind Energy Systems

Environmental impacts majorly include noise pollution, visual impairment, and ecological disturbances such as bird fatality and aquatic wildlife. An irritating sound occurs due to the mechanical and aerodynamic interaction of turbine with the wind. The aerodynamic noise is majorly caused due to the different velocity of wind moving through the turbine blades. The noise created through this interaction is so high that it can cause stone-deaf and sleep disorders for human beings [30]. This aerodynamic noise can be prevented by improving the aerodynamics of the blades. Thus, wind energy systems are required to be installed at a significant distance from the residential area. Besides noise pollution, visual impairment is the significant impact usually discussed. The sunlight reflects from the blades of the turbine and cause shadow flickering effect or disco effect. This effect creates disturbance in the visual capability of the nearby residential people which concludes as a negative machine for them [31].

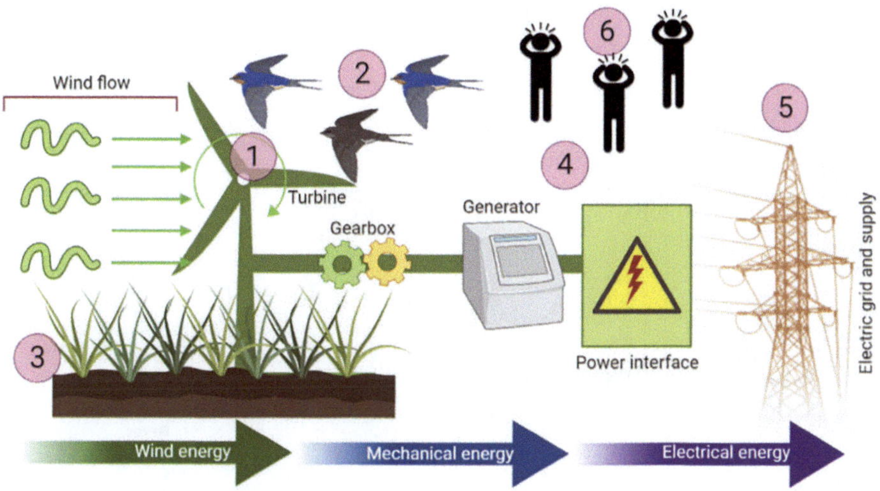

Fig. 5 Schematic of various environmental impacts of wind energy system [26]

As part of the ecosystem, the impact on animals and plants can never be neglected [32]. Birds are directly affected by the wind turbines, as they collide with the blades of the wind turbine and get killed. However, bird fatality due to wind turbines is significantly lower than other factors such as car collision, electricity, and human activities. According to the study, for every 250 cases of bird fatality due to human activities, only one bird dies due to wind turbine [33]. During the installation of a wind energy system, local climate is affected due to the removal of plants and vegetation from the land. Using large machinery, transportation of turbine and associated parts and also the connection of electric grid may lead to damage the fertility of soil and also cause soil erosion. The installation of wind turbine in the offshore areas leads to create turbulence on the surface of the ocean, which tends to decrease the amount of sunlight reaching the ocean. This effect causes harm to the aquatic life in the sea and may cause fatality of dolphins and seal as well [34] (Fig. 5).

2.4 *Hydro Energy Systems*

Hydro energy is the most ancient and mature energy source in the world. Hydro power plants are being used for almost a century and are still the most reliable energy source that exists [35]. Hydro energy power plant includes building a dam on a running river. The water from a higher altitude is released on turbine blades. The rotor of the turbine is coupled with a generator. The water rotates the blades of the turbine, this rotatory motion converts the mechanical energy into electrical energy using generator. Hydropower is the only renewable source that meets high demand of energy supply due to its 90% efficiency [36]. According to the Fig. 2. Hydro

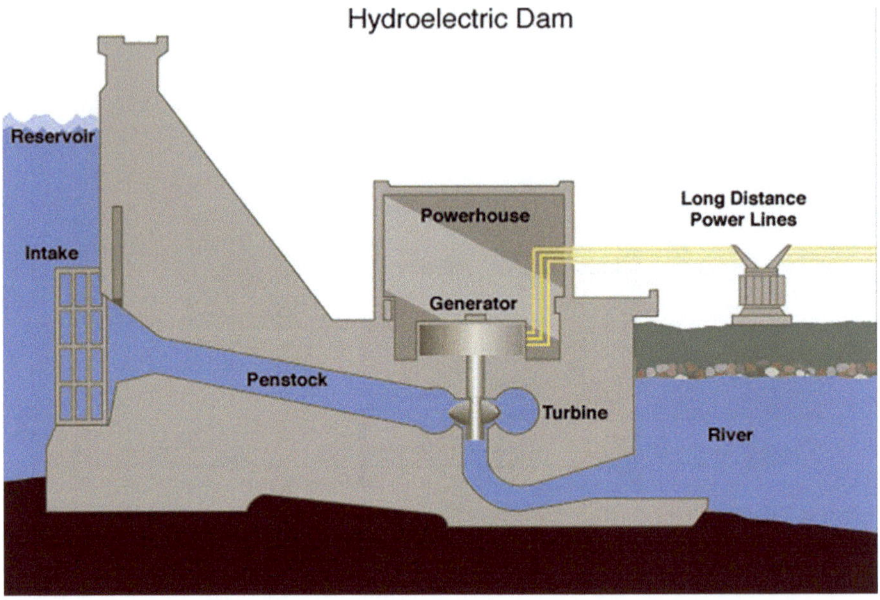

Fig. 6 Schematic of a conventional hydropower system [38]

energy shares 44.44% share of the renewable energy produced globally. On, the other hand hydropower technology has its own sociopolitical issues and ecological impacts those cannot be neglected [37] (Fig. 6).

2.4.1 Environmental Impacts of Hydro Energy Systems

Large-scale constructions are required for construction of a dam which results in high emission during the establishment of a dam [39]. A dam holds a large amount of water which acts as a barrier for the migration of fishes in the rivers. The conditions of flood caused by dams also disturbs the wildlife in that region. Dams alter the temperature of the nearby areas and changes the flow of rivers which further results in drying canals. High rate of evaporation is observed due to stored water in reservoir than running streams. This leads to dry out downstream of certain rivers that further affects the agricultural practices and animals [40]. The studies have shown that building up of large dams has resulted in increasing large amount of water on particular areas which has also resulted in a minor change in rotation of the sun. Thus, this impact of earth is a serious concern. Besides, relocation of the native people from the region to develop a hydro power project creates an uncomfortable living for them. Large amount of cultivation land and natural vegetation is destroyed, as the area occupied by the dam is high [41]. Dam is mainly constructed using concrete and steel which causes indirectly large emissions when transported through fossil fuels and constructed using fossil fuels [42].

2.5 *Biomass Energy Systems*

Biomass is one of the oldest forms of energy used form the ancient ages. For centuries, people have been using woods as the source of cooking food and heating their surroundings. Biomass is a primary source of food and energy [43]. Bioenergy has gathered the attention of scientists since last three decades and this technology is continuously improving. Combustion of biomass for the energy generation was the most conventional manner of handling the biomass which has a numerous impact on environment and human health, making it a negative source. Developing biomass power plants is the most feasible manner of handling the biomass rather than open dumping or combustion. Many technological processes have been developed to prevent the decay of biomass and convert to into high value products. These processes are gasification, pyrolysis, anaerobic digestion, torrefaction and other liquid biofuel production processes which can be found ecofriendly management of the biomass [44–47].

The abundance of biomass has increased its importance as an energy source. Lignocellulosic biomass such as food waste, wood, garden waste and farm waste, etc. cannot be preferred as food thus can be used as a raw material for power generation [48]. Certain studies have reported the use of biomass instead of coal in the conventional power plants so as the decrease the consumption of fossil fuels specially coal in the power plants to generate the energy sustainably [49]. Lignocellulosic biomass is a fibrous material comprising of cellulose, hemicellulose and lignin. Thus, is majorly being studied by scientists to produce biomaterials using biomass that can be compatible with the environment. These materials are either used individually or as a bio-composite with other materials causing waste-to-value scenarios [50] (Fig. 7).

2.5.1 Environmental Impacts of Biomass Energy Systems

Besides, various advantages and easy-to use potential of biomass there are certain environmental impacts associated with the biomass energy systems. Urban solid waste is generally burnt as a waste management practice as well as producing energy. Buring of this waste can lead to harmful chemical emissions into the environment [52]. Biofuel like ethanol and alcohol fuels have been found promising for automotive engines by various studies [53, 54]. However, production of these fuels affects the food supply chain system. Agricultural lands are being dedicated to the crops that can be used as a feedstock for biofuel production. This would result in the disturbance of food-energy nexus. Meeting the energy demand rather than food demand would lead to food scarcity and another crisis for human beings [55]. Converting forests into croplands for bioenergy would threaten the wildlife. A large amount of water is required to cultivate the non-edible crops for energy production, thus promoting clean water scarcity. Moreover, burning of these fuels produce emissions that cannot be controlled. Buring of biodiesel in diesel engines produces high amount of NOx, that

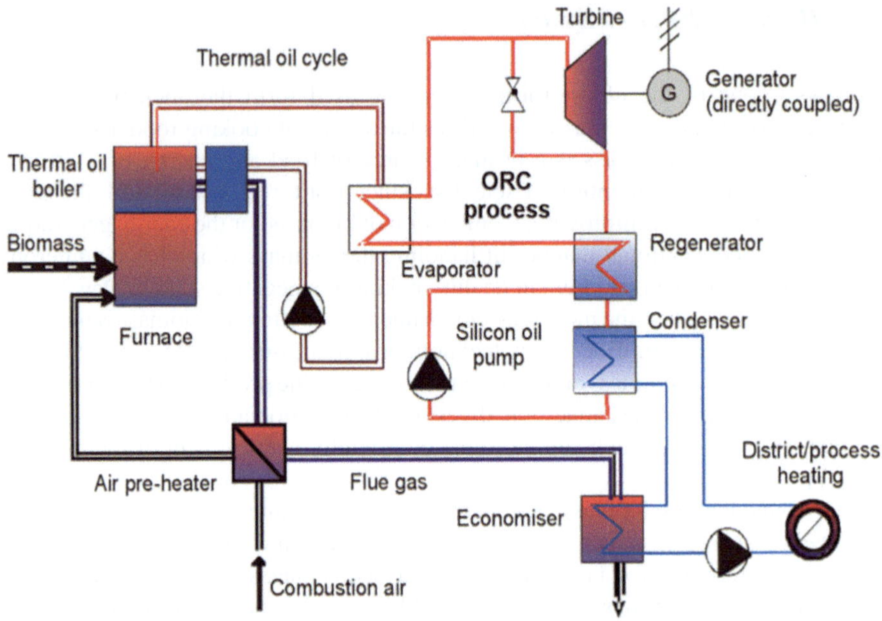

Fig. 7 Schematic of biomass energy system [51]

can result in harmful diseases if inhaled by humans. Other emissions like CO, CO_2, Sox and smog are also emitted but less than the amount in the case of conventional diesel [56]. Another challenge with the biomass energy is that it is not a zero-emission source of energy. A significant amount of harmful emissions are emitted whenever the energy is produced using raw biomass or processed biomass. Thus, making it a low promising energy source concerned to the environment.

2.6 SWOT Analysis

See Table 1.

Table 1 SWOT analysis of all energy systems

Energy Systems	Strength	Weakness	Opportunities	Threats	References
Solar energy	• Sun is the source of energy • Net operational emissions are zero • Small scale to large scale installation possible • Small business opportunities to generate economy • Affordable energy generating source	• Lifespan is around 25–30 years • The amount of output is uncertain due to solar power irradiation • Needs high amount of water for cleaning and cooling • Initial cost a implementing a solar plant is high • Large land area is required which disturbs the food-energy nexus	• Rooftops solar can be a major opportunity for renewable energy usage • Many cases of health issues like asthma and cardiovascular disorders can be avoided • Simple design and efficiency of solar panels can be improved significantly • Mining areas or floating solar panels can be opted for solar installation rather than agricultural lands	• Agricultural land occupied for solar power can affect the fertility of the soil • Toxic elements are produced after the lifespan of a solar panel • May cause eutrophication in nearby water bodies • The toxic waste can harm the human beings, birds and animals • Batteries are required for power storage which can be harmful for the environment after their lifespan	[57–61]

(continued)

Table 1 (continued)

Energy Systems	Strength	Weakness	Opportunities	Threats	References
Wind energy	• Second largest power generating source capacity in the world • No emissions are produced during operation • Recycling of wind turbines is simple • The turnover time of its carbon footprints is less than a month	• Visual disturbance • Noise pollution due to installation and low maintenance • Large accidents may occur due to storm • Flashing and flickering effect causes problem to the local people	• Can be installed both offshore and onshore • Can prevent both asthma attacks and other human diseases • Can prevent the consumption of water for power generation • Environmental impacts are reduced due to recycling • Improvement in design blades can be done to improve the noise problem	• Offshore installation causes disturbance to aquatic life • Can create headaches and sleeping disorders • Requires deforestation • Bird fatality • Causes soil erosion	[66–68],

(continued)

Table 1 (continued)

Energy Systems	Strength	Weakness	Opportunities	Threats	References
Hydro energy	• Most efficient renewable energy source with 90% efficiency • Oldest and most mature renewable energy technology • Generates 17% of the total global energy demand • Water is the raw material for generating power, which is abundant • Have high life span of 50–100 years	• Relocation of local residents is required, which causes discomfort • Change in local climate of the region, mainly increase in temperature • Dam causes hindrance in movement of aquatic life in the river • Steel and concrete required for construction of dam are produced by emission causing activities • Evaporation of water increase due to accumulation in a region • High cost of construction	• Cheap power generation • Hill areas should be preferred than flat regions. This provides proper height to the dam • Control flood and helps in irrigation in nearby areas • Minimal resources are required during operation	• Dried up rivers and canals cause loss in vegetation • Accumulation of water causes eutrophication • Dams may increase the erosion rate in rivers • Fish industries may face crisis due to hindrance in fish movement and reproduction activities • Natural wildlife may get extinct from the region	[62–65]

(continued)

Table 1 (continued)

Energy Systems	Strength	Weakness	Opportunities	Threats	References
Biomass energy	• Produces electricity by various processes • Greenhouse gas-carbon dioxide is consumed in the process • Easy to install small power plants • Both portable and immovable power producing plants can be used	• Burning biomass directly can produce harmful gases • Unpleasant smell • Agricultural land used for growing energy crops increases food scarcity • Large amount of water is required to cultivate the crops • Emissions are produced during biomass energy generation and consumption	• Existing coal plants can use biomass directly or in blended form with coal as source material • Biomass Energy is feasible to store for a long time • Improvement in biomass energy systems to increase efficiency and decrease emissions	• Huge amount of land and water is used • Causes deforestation due to converting forest land to cropland • Causes soil erosion • Municipal solid waste can emit harmful emissions if burnt as fuel • Lack of agricultural land for food can cause food scarcity	[57–72]

3 Role of AI

3.1 Role of AI in the RE Systems

The advent of AI can be traced back to 1954, credited to a pioneering computer scientist named McCarthy. Since its inception, AI has found applications across numerous diverse sectors, including marketing, banking, agriculture, healthcare, security, robotics, speech recognition, chatbots, and manufacturing, among others. During the past decade, AI has gained increasing attention in the area of energy systems, especially the energy grid, resulting in an evolution of the way in which energy is generated, distributed, and consumed [73]. The design existing power grid is insufficient to manage the integration of RE systems (e.g., wind, solar, geothermal, hydrogen). The unpredictability and variable nature RE systems create challenges in meeting the varying load of the power grid. The application of AI and DL (Deep Learning) models has shown potential to enhance the forecasting and planning of energy generation and distribution (Fig. 8).

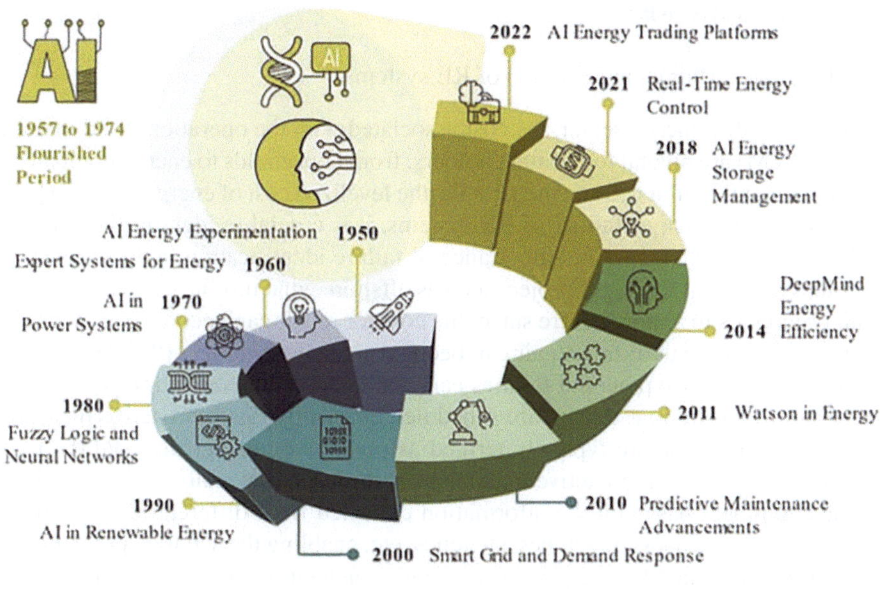

Fig. 8 Tentative overview of development of AI in the energy sector over the years [74]

3.2 Application of AI in RE Systems

In the realm of energy systems, AI finds applications in optimization, control, management, supply, and even policymaking. Over the years, its significance in the renewable energy sector has only amplified. While initially employed to sift through data and identify patterns for energy companies, AI can now be used to optimize the power generation, maintenance, and control of RES [75]. AI is now integrated in all the RE systems from wind, solar, hydro, ocean, geothermal to solar photovoltaic systems, among others. As a result, this integration of AI with the RE systems can effectively reduce costs related to balancing supply and demand, adjusting energy profiles, and managing grids for the energy sector [76] (Fig. 9).

The following are the different areas where the applications of AI in the renewable energy landscape have been transformative. Whether it's ensuring the longevity of renewable energy equipment or fine-tuning the energy production processes, AI stands at the forefront of innovation in the sector. Outlined below are some of the most significant applications of AI in renewable energy [78, 79].

3.2.1 Integrated RES

(a) Predictive Maintenance (PdM) of RE systems

In the renewable energy sector, the cost associated with the operational and maintenance (O&M) are pivotal for all stakeholders, from households to energy companies, and significantly influence key metrics like the levelized cost of energy (LCOE) [76]. To minimize the risk of failing of RE systems, it is crucial to identify the cause of failure in the early stage. The importance of failure identification in the early stage becomes important for larger project such as offshore wind turbine projects as once a fault progresses to a more severe stage, the cost for repair can become exponentially larger and the time until full repair can become extensively longer [80]. Hence, it is imperative to foresee potential failures early on to prevent severe damage to equipment and to ensure that repairs are scheduled efficiently. Such proactive measures for equipment in use are typically termed as predictive maintenance (PdM). PdM refers to a data-driven, proactive maintenance approach that analyzes the condition of the equipment based on the information collected from different manufacturing sources, such as sensors, machines, switches, etc. enabling the accurate prediction of optimal maintenance timing [78]. An essential element of predictive maintenance is Condition-Based Monitoring (CBM)—a systematic process that continuously observes the operational condition of critical components and uses real-time data by analyzing the equipment's sensor signals for failure forecasting. This approach allows for more efficient resource allocation, reduced downtime, and extended asset lifespan [81].

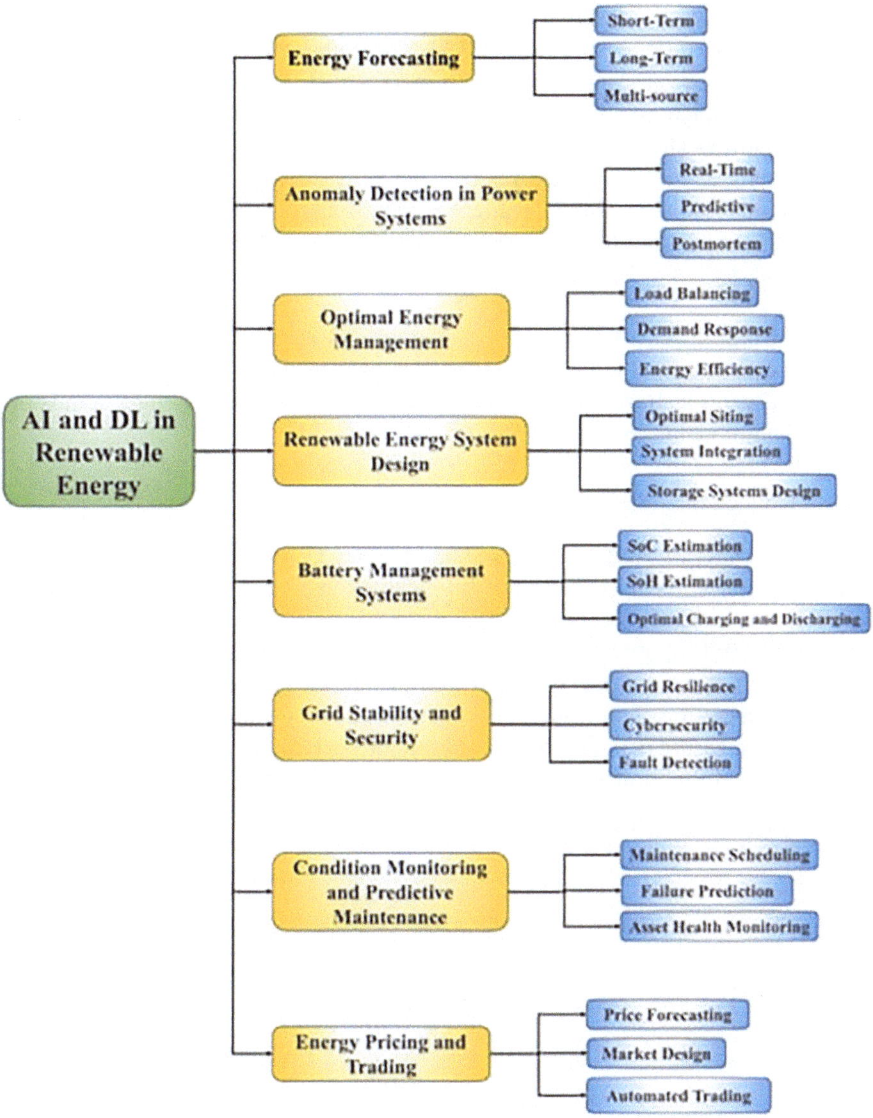

Fig. 9 The application of AI in different RE sectors [77]

What is Condition-Based Monitoring (CBM)?

CBM is an extended version of PdM where automatic triggering alarms are activated before obtaining any breakdown. The process to identify the failures and develop maintenance strategies starts with the Condition-Based Monitoring (CBM), which is a technique that involves tracking, collecting, and analyzing real-time data from renewable energy equipment, such as wind turbines or solar arrays, to assess the current state of their components [82]. The gathered data is subsequently pre-processed to remove any outliners. Hence, making it easier for the AI to interpret and facilitates the decision-making process. Lastly, the processed data is analyzed using machine learning algorithms, such as Artificial Neural Networks (ANNs), which forecast potential equipment failures by evaluating the machinery's current operational states. These algorithms are adept at identifying patterns within the data that may indicate impending failures or performance issues [76] (Fig. 10).

Step-by-Step Breakdown of CBM Using AI:

Below is a high-level overview of how data moves from machines and sensors to the AI models responsible for predictive maintenance. This figure outlines the core stages of Condition-Based Monitoring (CBM) using AI, bridging data acquisition and ongoing model refinement [83] (Fig. 11).

- Data Collection:

The effectiveness of predictive maintenance relies heavily on the availability and quality of data. The data used in PdM typically comes from two primary sources: historical maintenance records and real-time sensor data [84]. Historical data comprises records of previous maintenance procedures, operational logs, and failure reports. Real-time sensor data is gathered on an ongoing basis from machinery during its normal operation. For instance, sensors installed in turbines, generators, and similar parts measure parameters like temperature, vibration, pressure, and electrical output. This information is vital for predictive maintenance, as it delivers immediate insights into the equipment's status. By examining real-time sensor readings, AI models can spot initial indicators of wear, enabling maintenance to be arranged

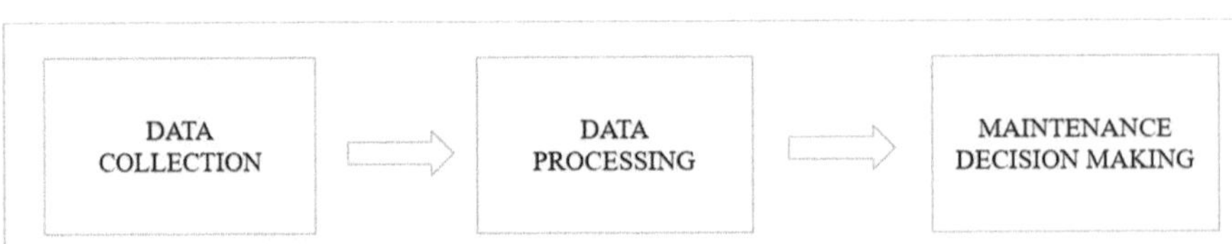

Fig. 10 Three steps for CBM

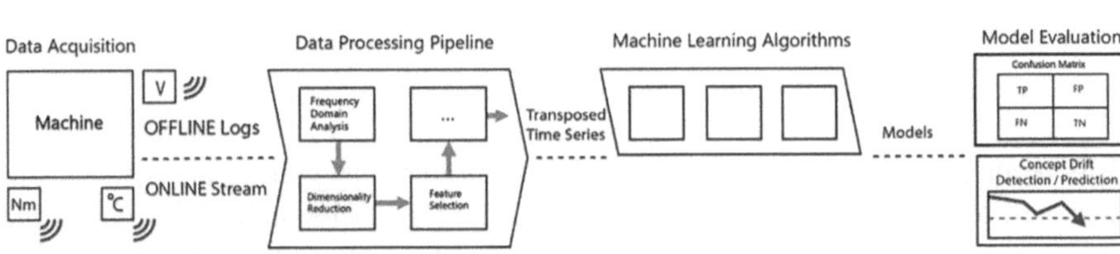

Fig. 11 Conceptual diagram of an AI-driven predictive maintenance pipeline [83]

before a breakdown happens. Such data is essential for training AI models because it establishes a baseline for identifying the patterns and trends linked to equipment failures.

- Data Pre-processing

This step is crucial prior to training and deploying as it helps to ensure that the data is both accurate and relevant. This step begins with data cleaning, where noisy or incomplete sensor logs are identified and addressed—often through techniques like interpolation for missing values [82]. Next, normalization and feature engineering help transform the raw signals (e.g., temperature, pressure, vibration) into consistent, meaningful indicators. Additionally, feature selection methods can highlight the most predictive variables, while reducing any redundancies that might slow down training. By refining the data at this stage, organizations enhance the reliability of predictive analytics, ultimately supporting more precise detection of potential faults and lowering the chance of unexpected equipment breakdowns [76].

- AI Analysis and Modeling

In this phase, advanced machine learning algorithm such as artificial neural networks (ANN) are trained on historical and real-time sensor readings to recognize the early warning signs of equipment failure. By learning the specific data trends and patterns linked with faults, the model can highlight subtle operational anomalies that indicate mechanical or electrical issues [85]. In practice, model selection often hinges on factors like computational efficiency and data complexity, while performance is validated against metrics such as accuracy, precision, and recall. This thorough approach to AI analysis and modeling ultimately increases the precision and dependability of predictive maintenance, minimizing unexpected failures in RES [82].

- Fault Detection

Building on the insights gained from AI modeling, fault detection aims to spot mechanical or electrical irregularities before they turn into major failures. Specifically, the AI algorithms trained on historical logs and live sensor data continuously compare current operating conditions against known "normal" patterns. When unexpected spikes in vibration frequency or temperature arise, the model flags these deviations as potential warning signs [86]. This early awareness is key to preventing unplanned downtime: technicians can promptly investigate and address the issue, often during a routine maintenance window. In wind power systems, for example, fault detection methods commonly track gear or blade deterioration, while in solar PV arrays, they look for rapid drops in output linked to panel degradation or soiling. By integrating automated alerts and operator review, fault detection mechanisms enhance overall reliability, reduce extensive repair costs, and safeguard the performance of renewable energy assets [87].

- Prediction and Decision-Making

Once the fault detection models flag an anomaly, prediction and decision-making frameworks come into play to determine the next best action for maintenance. These AI-driven modules forecast how a fault might evolve, assessing the likelihood and timing of a potential breakdown based on real-time and historical data. Equipped with these probabilistic insights, maintenance teams can schedule repairs at the most cost-effective intervals often aligning with off-peak production times or routine inspection cycles. Additionally, decision support tools can suggest parts replacement strategies or workload adjustments, ensuring that even minor issues are addressed before they escalate. By merging predictive analytics with strategic planning, renewable energy operators reduce unplanned downtime, prolong equipment life, and optimize overall operational efficiency [87].

The predictive capabilities that AI affords the renewable energy sector don't only benefit when it comes to supply and demand. It is also useful for the maintenance of infrastructure. For example, AI-powered predictive analysis collects data from wind turbine sensors, monitoring the infrastructure for wear and tear and alerting the operator when maintenance is needed. In addition to this, AI can assist human inspectors by providing diagnostic suggestion considering the machine's current condition, the operating environment, and the history of similar breakdowns and repairs [88]. This assistance is expected to enhance predictive maintenance practice by overcoming the limitation associated with human inspection process such as fatigue, and variability [88]. Hence, the AI can not only reduce the frequency of the unplanned or corrective maintenance, but it can also alert the system by highlighting abnormal behaviors detected in the system prior to any failure by constantly monitoring the machine's current operating state. This implementation in the RE systems can result in saving O&M costs, reduced machine down-time, and improving supply stability. It can also result in the prolonged operation of the equipment, and optimum operation of the of the RE systems.

3.2.2 Energy Generation

As the attention towards clean and sustainable energy is increasing, research on renewable energy technologies has spiked. The RES have shown potential in reducing the energy dependence on foreign countries and producing energy at lower costs. However, the variability and intermittent have limited the potential of RE systems. For example, solar energy is influenced by the cloud formation and variation in sun-hours due to seasonal changes, while the wind energy heavily depends on the optimum wind speeds, and any changes in the weather can heavily influence the energy production [89]. The accurate weather forecasting will not only lead to increased energy generation, but will also help in site selection, installation, managing supply and demand, routine maintenance, and decommissioning sites. Moreover, as accurate weather forecasting is crucial for maximizing energy production and improving grid integration and stability, precise demand forecasting becomes equally important for predicting system energy requirements over time. In addition, accurate weather forecasting can also enable the safe planning of future routine operation and maintenance works, especially for offshore structures [90]. Demand forecasting is important for economic and safe operations planning in power distribution systems [91]. Hence, developing an accurate real-time forecasting system for both weather and load is crucial, which can help in reducing the negative impact of the variability and uncertainty of renewable energy generation on the power grid, and helps in optimizing the energy production and distribution without overloading the RE systems and existing grid.

The forecasting can be performed according to the different lead times and horizons, which can be roughly categorized into very-short term, short-term, medium-term, long-term (as shown in the table below). This multi-tiered forecasting strategy can enable energy traders to make informed decisions that are aligned with both immediate operational needs and long-term strategic goals. By anticipating fluctuations in energy production and demand, utilities can optimize resource allocation, minimize costs, and plan maintenance effectively. This is essential for maintaining a balance between supply and demand, which is critical for the stability of the energy grid and for meeting the challenges of a dynamic energy market [92].

(a) Weather and Demand Forecasting

For years, energy generation has been predicted using traditional forecasting models. These models include statistical models such as the Autoregressive Integrated Moving Average (ARIMA) and Seasonal ARIMA (SARIMA), which have been widely applied due to their robustness in capturing linear relationships in time series data. Their limitations involve their inability to handle complex nonlinear relationships and the high-dimensional nature of renewable energy data. Additionally, numerical models such as, Numerical Weather Prediction (NWP) models, which utilizes mathematical representations of the atmosphere and numerical methods to predict wind speed and direction, while solar radiation models use cloud cover and atmospheric conditions to predict solar irradiance based on current conditions [91]. In the present day, NWP is considered to be one of the most accurate weathers forecast systems used by the National Weather Service's Global Forecast System, or GFS, and the European Center for Medium-Range Weather Forecast (ECMWF). They are also known as the American and European models, respectively. In general, the European model has produced the most accurate forecasts of global weather [93]. However, their limitation include uncertainty in weather forecasting due to non-linear and complex patterns associated with weather data, which are crucial for predicting the variable and intermittent nature of renewable energy sources like wind and solar power [94]. Nowadays, with the advent of big data and advancements in computational power, AI-implemented weather forecasting models are increasingly being adopted. Machine Learning (ML) and Deep Learning (DL) algorithms can learn complex nonlinear relationships from vast amounts of data, making them suitable for handling the high-dimensional nature of renewable energy data [94]. AI programs combine 'self-learning' weather models, historical weather datasets and real-time local weather information to predict power generation and communicate with grid capacity levels. The combination of ML algorithms such as Support Vector Machines (SVM), Support Vector Regression (SVR), Convolutional Neural Networks (CNN), and ANN with traditional forecasting models has shown enhanced accuracy in predicting solar and wind energy parameters such as solar radiation and wind speed. These are referred to as Hybrid Models (HM) [95]. The ANN based model achieved the most accurate results for predicting daily global solar radiation data, while the SVM based models are superior to other model for wind speed prediction [94]. Moreover, the ANN based model is more accurate for the forecasting the electricity consumption of different sectors of industrial, residential, non-industrial, commercial, public lighting and entertainment [96].

The integration of AI techniques with traditional forecasting models heralds a new era in renewable energy forecasting, enhancing the reliability and precision of renewable energy forecasts, which is imperative for the sustainable management and integration of renewable resources in our energy systems. AI's advanced data processing capabilities enable the bridging of gaps between model predictions and actual conditions, resulting in more reliable forecasts. The vast amounts of data from satellite observations, when processed through AI, reveal new patterns and models, reducing bias and enhancing forecast precision. While AI offers numerous advantages such as speed in prediction execution, it is not without its limitations. The accuracy of AI models is heavily dependent on the quality and volume of data available, and there remains an inherent challenge in the unpredictability of weather patterns and climate phenomena. Moreover, while AI can significantly improve the efficiency of spatial land planning and the sustainability of land use, the complexity of ecological systems and the need for extensive datasets pose further challenges. Nevertheless, as AI technologies continue to evolve and the reservoir of data expands, we can anticipate a future where AI not only informs but also transforms our approach to managing the effects of climate on human life and the environment [79].

3.2.3 Energy Distribution

Due to technological advancements, the demand for electrical energy is rising, that has not only become a challenge for its production but also its distribution. Although the energy-efficiency legislations and the efficiency of energy end-use services have progressed gradually. It is not enough to off-set the growth in energy services, such as commodity production and consumption. This rise in energy demand is growing the complexities associated with the traditional power grid system by increasing requirement for greater reliability, efficiency, security, and environmental and energy sustainability concerns [97]. To maintain stability within the grid, it is crucial for the energy generated to match the energy consumed precisely. The variability and uncertainty associated with the RE systems cause disruptions in the power supply, which leads to fluctuations in the supply and demand of energy for a particular place can cause instability in the grids. The traditional grid, being unidirectional, simply facilitates energy flow from generation facilities to end-users, without the capability for real-time adjustments or two-way communication [98].

In addition, traditional grids face the problem of energy theft, which compromises the integrity of the grid and results in substantial revenue losses. Energy theft, often resulting from unauthorized access or tampering with infrastructure, exacerbates the difficulty of managing demand-response accurately. Conventional power distribution systems rely on the phenomenon of remote generation, varying voltage for transmission, distribution, and consumption in accordance with the average demand in a given area, installing physical protection equipment at multiple nodes, and measuring monthly consumer consumption [99]. In the absence of sophisticated monitoring systems and a responsive communications network, aligning supply with fluctuating demand profiles becomes an impossible challenge [100].

In this context, the integration of AI into smart grid technology is seen as a promising solution. By enabling bidirectional communication, smart grids effectively balance supply and demand for energy by monitoring and controlling energy flows in real time. AI algorithms within smart grids can predict consumption patterns, detect anomalies indicative of energy theft, and autonomously re-route power to optimize grid performance, thereby enhancing security and sustainability. With AI added to the grid, it becomes a dynamic, self-regulating network that can handle the complexities of modern energy consumption.

(A) Smart Grid

A smart grid is an advanced electrical network that incorporates digital technology, sophisticated sensors, and information and communication technologies (ICTs) to improve the efficiency of electricity distribution. The smart grid using the combination of AI algorithms such as convolutional neural network (CNN) and a long short-term memory (LSTM) architecture, provides the utilities and customers' facility to monitor, control, and predict energy use by integrating modern digital equipment with the existing electrical system [100]. AI algorithms can predict energy demand patterns by analyzing historical and real-time energy consumption data, identifying inefficiencies, and adjusting energy generation and distribution dynamically. In addition, AI-driven analytics enable utilities to identify and respond to grid disturbances or anomalies, thereby increasing grid resilience and reducing downtime [77] (Fig. 12).

According to the U.S. Department of Energy, a smart grid is defined as "A smart grid uses digital technologies to improve the reliability, security, and efficiency (both economic and energy) of the electric system-from large-scale generation to utility systems to electricity consumers and a growing number of distributed generation and storage resources" [102] (Fig. 13).

It is a bidirectional flow of energy and information that allows the utility to efficiently manage the transmission and distribution of energy while providing customers with the ability to influence energy consumption [102]. This two-way communication between the consumer and the utility is possible because of a crucial smart grid technology, known as smart meter. The smart grid functions as an Internet of Things (IoT) network implementations with smart meters and wireless smart sensors placed

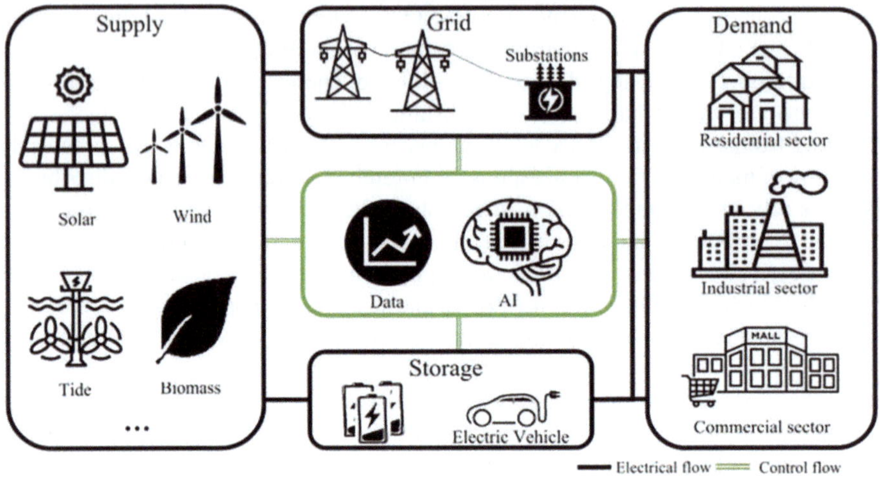

Fig.12 A brief overview of Smart Grid and its components [101]

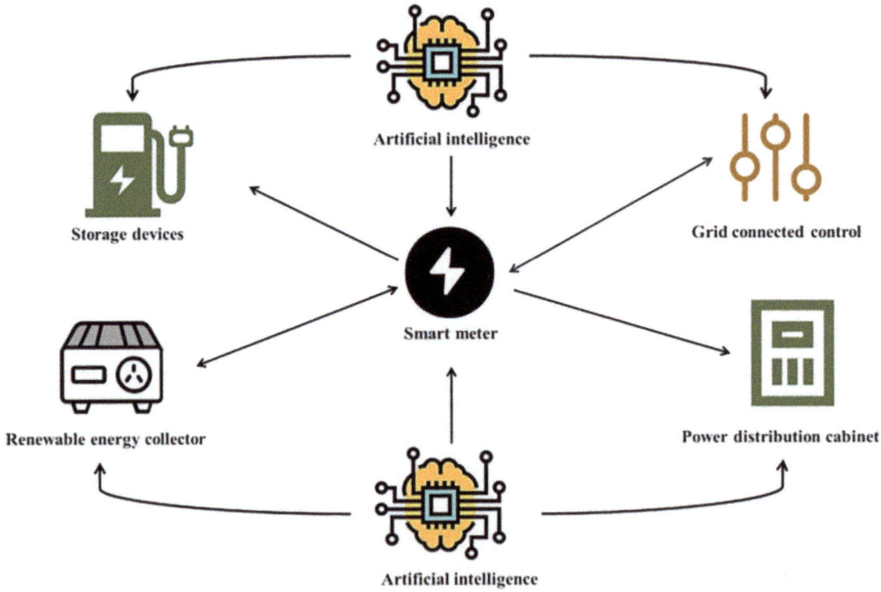

Fig. 13 Smart meters and its components [79]

throughout the grid that provides utilities and consumers with two-way communication of electricity and information, making it an integral part of the generation of sustainable and intelligent energy [102, 103]. It optimizes energy flows by dynamically aligning energy supply with demand in real time, utilizing connected devices and real-time data analytics. In addition to reducing expenses and optimizing energy

Table 2 Technical areas governed by the implementation of smart grid

Reliability	– Customer services improved – Good power quality – Adequate outage frequency and duration
Security	– Protection against cyber security attacks – Protection against electricity theft – Protection against natural faults
Economics	– Reduced electricity bills – There would be an option for consumers to save electricity – Reduction in maintenance and operational costs
Efficiency	– Options for energy conservation are available – The intensity of system losses will reduce
Environmental	– Lower emissions due to greater use of renewable sources – Not as much dependent on the weather conditions
Safety	– Protection for the line workers – Protection for the public

efficiency, the networked system seeks to increase the stability and dependability of the grid. The application of smart grid basically brigs improvement in several key technical area as described in the Table 2 [104].

Smart grid operators, recognizing AI's potential, have leveraged its capabilities to create substantial resource benefits. Despite these advancements, AI's reliance on extensive historical data and the need for continuous data updates present notable limitations. The technology requires ongoing refinement to keep pace with the evolving energy landscape and to ensure the predictive models remain accurate and effective. As AI continues to mature, its role in driving sustainable energy transitions and in facilitating more intelligent, reliable, and efficient power grids becomes increasingly indispensable.

3.2.4 Renewable Energy Storage

The renewable energy sector has emerged as a vital solution in reducing the effects of global warming and climate change by limiting global greenhouse gas emissions. Renewable energy harnesses natural resources such as sunlight, wind, water, geothermal, heat and biomass to generate energy, and offers a cleaner and more sustainable alternative to traditional fossil fuels [105]. However, intermittent and variability are the biggest limitations associated with renewable energy technology. The RE systems heavily depend on the availability of these natural resources to generate energy and are subject to fluctuation due to environmental conditions or time of day. The inherent variability of this renewable energy source causes intermittent power generation, which makes it difficult to match supply with demand consistently. Due to the intermittent nature of renewable energy, disruption is caused in continuous energy distribution, which can lead to energy waste, grid instability, voltage fluctuation and system outages. Moreover, there can be times when the energy produced is either insufficient to meet demand or so excessive that it cannot be fully utilized. An effective energy storage system is crucial to overcoming these challenges [106]. Energy Storage refers to the capturing and storing of excess energy generated during periods of low demand and making it available when demand is high or renewable sources are intermittent [107]. There are various ways to store energy such as mechanical, electro-chemical, thermal energy storage etc. with electro-chemical being the most efficient and widely used energy storage system.

AI can significantly enhance the efficiency of energy storage systems in managing the variability of renewable energy sources. AI algorithms can forecast energy production and predict energy surges and troughs by analyzing vast datasets, including weather patterns, historical energy usage, and grid performance. In other words, energy providers can save energy that can be distributed more effectively for commercial or residential uses at the time of peak demand. This integration of AI with the energy storage is referred to as 'Intelligent Storage' [108]. For example, machine learning models can be trained to predict solar irradiance and wind speed to estimate the output of solar panels and wind turbines. This information allows the energy storage systems to be controlled proactively, reducing the need for more costly and carbon extensive power plants that are traditionally used to manage demand spikes.

Moreover, based on these predictions, AI can optimize the charging and discharging cycles of energy storage systems to ensure that energy is available when needed, thus improving grid stability and the integration of renewable energy sources [89]. Electro-chemical storage systems have higher energy densities, and can achieve a wide range of round-trip efficiency, as high as 95% for Li-ion batteries [109]. The use of batteries is increasing rapidly for residential energy storage and electric vehicles. The intelligent battery system can optimize the charging and discharging of energy storage equipment, which directly influences its life span. By continuously monitoring fundamental parameters such as current, voltage, temperature, and even strain and pressure through advanced sensing technologies, these AI-enabled BMS can diagnose the state of health (SOH) and predict the ageing trajectory of each battery cell in real-time [110]. These applications enable diagnosing each battery cell every minute with real-time data and with the assistance of data visualization [89]. The extensive research in AI-based battery health management underscores the importance of these strategies for the effective operation and longevity of battery systems. The figure describes the implementation of AI in the life cycle of the battery system, from battery manufacturing to the operation stage. AI implementation with energy storage systems represents a transformative step in harnessing the full potential of renewable energy sources. By mitigating the inherent intermittent and variability of renewables such as solar and wind, intelligent storage systems equipped with AI significantly enhance the stability and efficiency of the energy grid.

4 Limitation

The integration of AI with RES is revolutionizing the energy sector, leading to substantial advancements in energy generation, distribution, and storage. AI algorithms have transformed standalone grid operations by enabling energy companies to conduct real-time monitoring and forecasting. These capabilities are essential for adapting to fluctuating demand and varying weather conditions, which are particularly relevant for RE systems like solar and wind power, where energy output heavily depends on environmental factors. One of the key innovations facilitated by AI is the development of smart grids. These grids utilize AI-driven analytics to optimize the flow of electricity, enhancing the efficiency of the entire energy network. Furthermore, AI significantly contributes to the advancement of intelligent storage solutions. This approach not only stabilizes the grid but also ensures a more reliable energy supply, mitigating the risk of blackouts and power outages. These technologies achieve this by effectively balancing demand and supply both nationally and locally [76]. However, relying solely on AI to manage these systems comes with its own set of limitations.

4.1 Handling Large Datasets

The training of AI models necessitates extensive datasets to ensure accurate performance in their designated tasks. The collection and labelling of such large dataset are a formidable challenge, often involving significant time and effort [76]. There is considerable importance to this process since the quality and diversity of the data directly impact the effectiveness and reliability of the model.

In many sectors, particularly those involving complex phenomena like weather forecasting or peak-demand load planning, gathering comprehensive data sets can be especially challenging. The challenge is further complicated when dealing with unstructured data, such as images or text, where labeling becomes even more complex. Despite these challenges, the effort invested in creating robust datasets is invaluable. A well-curated data set can significantly enhance the performance of AI systems, allowing them to make more accurate predictions, identify patterns more accurately, and provide more valuable insights [111]. The development of more efficient data collection and labeling methods is becoming an important area of focus as AI advances, with the potential to streamline this process and enhance the capabilities of AI systems across a wide range of disciplines.

4.2 Integration with Existing Infrastructure

Undoubtedly, the integration of AI with existing power grid systems harbors immense potential to transform them into more optimized, secured, and resilient operations. However, incorporating AI into traditional power grid systems presents multiple challenges that complicate this technologically progressive endeavor. A primary obstacle is data management: the energy sector generates extensive data, such as from smart-meter rollouts, but this data often lacks the real-time, centralized preparation necessary for effective AI utilization. The inherent complexity of the energy sector, compounded by dense regulatory frameworks, further complicates AI integration. This complexity is particularly evident in the heterogeneous and decentralized nature of modern power grids, especially at lower and medium-voltage levels, where effective monitoring becomes challenging [112].

Additionally, integrating AI into power grids raises significant cybersecurity concerns. Given the critical importance of maintaining grid integrity, there is a pressing need for secure data transfer technologies and reliable parameters to ensure system security [113]. The path forward involves a focused approach on early data recording, storage, the development of sophisticated AI applications for load and supply predictions, and a cross-industry learning approach to leverage advanced digital practices from sectors like logistics.

4.3 Cybersecurity

The AI heavily relies on technologically advanced sensors and data collection instruments, interconnected in a network commonly known as the Internet of Things (IoT). As digital technologies become more prevalent, the need to connect previously isolated components with larger communication networks increases, leading to many new data flows. When IoT is integrated with the power grid, it forms an extensive cyber-physical system network capable of remotely monitoring and controlling connected devices [99]. Traditional grid faults, often associated with equipment failure, can be easily overcome. However, faults in the Smart Grid (SG) are predominantly digital, such as cyber-attacks or data transfer issues. Other attacks include injecting false information, energy thefts, disconnecting resources, and introducing malware, causing blackouts or power outages. Indeed, cyber-attacks on power grids have already succeeded in causing temporary, large-scale blackouts in the recent past [114].

Many advanced AI models have been developed as part of the latest research to overcome these challenges. These models have proven effective in detecting grid failures, such as cyber-attacks, and in handling them, as well as in predicting potential failures. Moreover, these models are capable of enhancing the computational capacity of the system while handling large data sets [115].

4.4 Regulations and Policies

The integration of AI with RES necessitates additional investments and financial reserves for technology transfer, infrastructure provision, communication systems, hiring skilled professionals (engineers and others), and research and development. However, the return on investment may be prolonged due to high initial costs [116]. Advanced AI technology requires massive computational processing, entailing extensive data analysis and monitoring facilities. These facilities have high energy requirements and carbon footprints, potentially compromising the outcomes of Sustainable Development Goals (SDGs) 7 and 13, which focus on affordable and clean energy and climate action, respectively [117]. There is a notable lack of infrastructure investment in research and development to facilitate greener and more sustainable integration of AI with the renewable energy sector. Another major concern is the collection of private data for training AI models, which raises issues of privacy threats and potential data misuse. The European Union has established regulations for the protection of natural persons in the processing of personal data and the free movement of such data [118]. However, this regulation has not been adopted by many other countries, posing a risk to minorities in regions where democratic control is lacking [117].

In addition to the aforementioned challenges of integrating AI with renewable energy systems, it is crucial to consider the ethical implications. Ensuring equitable access to energy resources and avoiding biases in AI algorithms is essential for fair distribution of energy services. Furthermore, the interoperability and standardization of systems across different regions and technologies are vital for seamless global integration. Public perception and acceptance also play a significant role; thus, targeted education and awareness campaigns are necessary to foster understanding and support for these technologies. Collaboration between various stakeholders—governments, industry, academia, and civil society—is imperative to address these challenges effectively and align AI integration with broader societal goals [119]. Lastly, future policy directions should focus on supporting green AI technologies and reinforcing data privacy and security, providing a framework that nurtures sustainable and responsible development in this field.

5 Conclusion

This book chapter has highlighted the transformative potential of AI in revolutionizing the integration with renewable energy systems. AI's capabilities extend to optimizing grid infrastructure, enhancing its resilience, and enabling real-time monitoring of demand and energy consumption. Notably, AI models have demonstrated proficiency in predicting system breakdowns in RE systems and forecasting weather patterns and demand responses. The chapter provides a comprehensive overview of conventional renewable energy technologies, including solar, wind, hydro, geothermal, and biomass. It delves into the benefits of integrating AI with these systems, particularly in addressing their inherent unpredictability and variability, which are major challenges in the sector. The chapter concludes by discussing the significant challenges that impede AI integration with RE systems and proposes future research directions. These include developing more advanced AI algorithms for better prediction accuracy, enhancing cybersecurity measures to protect grid integrity, and fostering interdisciplinary collaboration to accelerate the transition towards a more efficient and sustainable energy future. By addressing these challenges, the integration of AI with renewable energy systems can be significantly improved, hastening the transition towards a more sustainable and resilient energy landscape.

References

1. Majeed, Y., Khan, M. U., Waseem, M., Zahid, U., Mahmood, F., Majeed, F., Sultan, M., & Raza, A. (2023). Renewable energy as an alternative source for energy management in agriculture. *Energy Reports, 1*(10), 344–359. https://doi.org/10.1016/J.EGYR.2023.06.032
2. Eamus, D., & Jarvis, P. G. (2004). The direct effects of increase in the global atmospheric CO_2 concentration on natural and commercial temperate trees and forests. *Advances in Ecological Research, 34*, 1–58. https://doi.org/10.1016/S0065-2504(03)34001-2
3. Salman, M., Long, X., Wang, G., & Zha, D. (2022). Paris climate agreement and global environmental efficiency: New evidence from fuzzy regression discontinuity design. *Energy Policy, 168*, 113128. https://doi.org/10.1016/J.ENPOL.2022.113128
4. Noumba, I., & Nguea, S. M. (2023). Assessing the role of globalization for universal electricity access. *International Economics, 174*, 180–195. https://doi.org/10.1016/J.INTECO.2023.03.008
5. Access to Electricity [Online]. Available: https://unstats.un.org/unsd/methodology/m49/
6. Davis, M., Moronkeji, A., Ahiduzzaman, M., & Kumar, A. (2020). Assessment of renewable energy transition pathways for a fossil fuel-dependent electricity-producing jurisdiction. *Energy for Sustainable Development, 59*, 243–261. https://doi.org/10.1016/J.ESD.2020.10.011
7. World Energy Outlook 2022 [Online]. Available: www.iea.org/t&c/
8. Rahman, A., Farrok, O., & Haque, M. M. (2022). Environmental impact of renewable energy source based electrical power plants: Solar, wind, hydroelectric, biomass, geothermal, tidal, ocean, and osmotic. *Renewable and Sustainable Energy Reviews, 161*, 112279. https://doi.org/10.1016/J.RSER.2022.112279
9. Sayed, E. T., et al. (2021). A critical review on environmental impacts of renewable energy systems and mitigation strategies: Wind, hydro, biomass and geothermal. *Science of The Total Environment, 766*, 144505. https://doi.org/10.1016/J.SCITOTENV.2020.144505
10. "gsr_2020_full_report_en"
11. Bist, N., Sircar, A., & Yadav, K. (2020). Holistic review of hybrid renewable energy in circular economy for valorization and management. *Environmental Technology and Innovation, 20*, 101054. https://doi.org/10.1016/J.ETI.2020.101054
12. Lee, J., Lin, K. Y. A., Jung, S., & Kwon, E. E. (2023). Hybrid renewable energy systems involving thermochemical conversion process for waste-to-energy strategy. *Chemical Engineering Journal, 452*, 139218. https://doi.org/10.1016/J.CEJ.2022.139218
13. Sorrenti, I., Harild Rasmussen, T. B., You, S., & Wu, Q. (2022). The role of power-to-X in hybrid renewable energy systems: A comprehensive review. *Renewable and Sustainable Energy Reviews, 165*, 112380. https://doi.org/10.1016/J.RSER.2022.112380

14. Basnet, S., Deschinkel, K., Le Moyne, L., & Cécile Péra, M. (2023). A review on recent standalone and grid integrated hybrid renewable energy systems: System optimization and energy management strategies. *Renewable Energy Focus, 46*, 103–125. https://doi.org/10.1016/J.REF.2023.06.001
15. Mellit, A., & Kalogirou, S. A. (2008). Artificial intelligence techniques for photovoltaic applications: A review. *Progress in Energy and Combustion Science, 34*(5), 574–632. https://doi.org/10.1016/J.PECS.2008.01.001
16. Aghbashlo, M., et al. (2021). Machine learning technology in biodiesel research: A review. *Progress in Energy and Combustion Science, 85*, 100904. https://doi.org/10.1016/J.PECS.2021.100904
17. Inman, R. H., Pedro, H. T. C., & Coimbra, C. F. M. (2013). Solar forecasting methods for renewable energy integration. *Progress in Energy and Combustion Science, 39*(6), 535–576. https://doi.org/10.1016/J.PECS.2013.06.002
18. Thaker, S., Oni, A. O., Gemechu, E., & Kumar, A. (2019). Evaluating energy and greenhouse gas emission footprints of thermal energy storage systems for concentrated solar power applications. *J Energy Storage, 26*, 100992. https://doi.org/10.1016/J.EST.2019.100992
19. Hosenuzzaman, M., Rahim, N. A., Selvaraj, J., Hasanuzzaman, M., Malek, A. B. M. A., & Nahar, A. (2015). Global prospects, progress, policies, and environmental impact of solar photovoltaic power generation. *Renewable and Sustainable Energy Reviews, 41*, 284–297. https://doi.org/10.1016/J.RSER.2014.08.046
20. Mishra, R., et al. (2023). Dynamics of solar energy entrepreneurship in rural Bihar, India. *Energy for Sustainable Development, 76*, 101269. https://doi.org/10.1016/J.ESD.2023.101269
21. Algarni, S., Tirth, V., Alqahtani, T., Alshehery, S., & Kshirsagar, P. (2023). Contribution of renewable energy sources to the environmental impacts and economic benefits for sustainable development. *Sustainable Energy Technologies and Assessments, 56*, 103098. https://doi.org/10.1016/j.seta.2023.103098
22. Oteng, D., Zuo, J., & Sharifi, E. (2022). Environmental emissions influencing solar photovoltaic waste management in Australia: An optimised system network of waste collection facilities. *Journal of Environmental Management, 314*, 115007. https://doi.org/10.1016/J.JENVMAN.2022.115007
23. Balali, M. H., Nouri, N., Omrani, E., Nasiri, A., & Otieno, W. (2017). An overview of the environmental, economic, and material developments of the solar and wind sources coupled with the energy storage systems. *International Journal of Energy Research, 41*(14), 1948–1962. https://doi.org/10.1002/er.3755
24. Sokka, L., et al. (2016). Environmental impacts of the national renewable energy targets—A case study from Finland. *Renewable and Sustainable Energy Reviews, 59*, 1599–1610. https://doi.org/10.1016/j.rser.2015.12.005
25. Sah, D., Chitra, & Kumar, S. (2023). Investigation and recovery of copper from waste silicon solar module. *Materials Chemistry and Physics, 296*, 127205. https://doi.org/10.1016/J.MATCHEMPHYS.2022.127205
26. Nazir, M. S., Ali, N., Bilal, M., & Iqbal, H. M. N. (2020). Potential environmental impacts of wind energy development: A global perspective. *Current Opinion in Environmental Science & Health, 13*, 85–90. https://doi.org/10.1016/J.COESH.2020.01.002
27. Mahmoud, M., Ramadan, M., Abdelkareem, M. A., & Olabi, A. G. (2023). Applications of wind energy. *Renewable Energy—Volume 1: Solar, Wind, and Hydropower Definitions, Developments, Applications, Case Studies, and Modelling and Simulation, 1*, 331–345. https://doi.org/10.1016/B978-0-323-99568-9.00010-8
28. Kuşkaya, S., & Bilgili, F. (2020). The wind energy-greenhouse gas nexus: The wavelet-partial wavelet coherence model approach. *Journal of Cleaner Production, 245*, 118872. https://doi.org/10.1016/J.JCLEPRO.2019.118872
29. Mahmoud, M., Ramadan, M., Olabi, A. G., Pullen, K., & Naher, S. (2020). A review of mechanical energy storage systems combined with wind and solar applications. *Energy Conversion and Management, 210*, 112670. https://doi.org/10.1016/J.ENCONMAN.2020.112670

30. Karasmanaki, E. (2022). Is it safe to live near wind turbines? Reviewing the impacts of wind turbine noise. *Energy for Sustainable Development, 69*, 87–102. https://doi.org/10.1016/J.ESD.2022.05.012
31. Haac, R., Darlow, R., Kaliski, K., Rand, J., & Hoen, B. (2022). In the shadow of wind energy: Predicting community exposure and annoyance to wind turbine shadow flicker in the United States. *Energy Research & Social Science, 87*, 102471. https://doi.org/10.1016/j.erss.2021.102471
32. Teff-Seker, Y., Berger-Tal, O., Lehnardt, Y., & Teschner, N. (2022). Noise pollution from wind turbines and its effects on wildlife: A cross-national analysis of current policies and planning regulations. *Renewable and Sustainable Energy Reviews, 168*, 112801. https://doi.org/10.1016/J.RSER.2022.112801
33. Wang, S., Wang, S., & Smith, P. (2015). Ecological impacts of wind farms on birds: Questions, hypotheses, and research needs. *Renewable and Sustainable Energy Reviews, 44*, 599–607. https://doi.org/10.1016/J.RSER.2015.01.031
34. Msigwa, G., Ighalo, J. O., & Yap, P. S. (2022). Considerations on environmental, economic, and energy impacts of wind energy generation: Projections towards sustainability initiatives. *Science of The Total Environment, 849*, Article 157755. https://doi.org/10.1016/J.SCITOTENV.2022.157755
35. Nazir, M. S., Ali, Z. M., Bilal, M., Sohail, H. M., & Iqbal, H. M. N. (2020). Environmental impacts and risk factors of renewable energy paradigm—A review. *Environmental Science and Pollution Research, 27*(27), 33516–33526. https://doi.org/10.1007/s11356-020-09751-8
36. Javed, M. S., Ma, T., Jurasz, J., & Amin, M. Y. (2020). Solar and wind power generation systems with pumped hydro storage: Review and future perspectives. *Renewable Energy, 148*, 176–192. https://doi.org/10.1016/J.RENENE.2019.11.157
37. Kuriqi, A., Pinheiro, A. N., Sordo-Ward, A., Bejarano, M. D., & Garrote, L. (2021). Ecological impacts of run-of-river hydropower plants—Current status and future prospects on the brink of energy transition. *Renewable and Sustainable Energy Reviews, 142*, 110833. https://doi.org/10.1016/J.RSER.2021.110833
38. El Bassam, N., Maegaard, P., & Schlichting, M. L. (2013). Hydropower. In *Distributed renewable energies for off-grid communities*, pp. 167–174. https://doi.org/10.1016/B978-0-12-397178-4.00010-4
39. Gagnon, L., & Van De Vate, J. F. (1997). Greenhouse gas emissions from hydropower: The state of research in 1996. *Energy Policy, 25*(1), 7–13. https://doi.org/10.1016/S0301-4215(96)00125-5
40. Inhaber, H. (2004). Water use in renewable and conventional electricity production. *Energy Sources, 26*(3), 309–322. https://doi.org/10.1080/00908310490266698
41. Fearnside, P. M. (2016). Environmental and social impacts of hydroelectric dams in Brazilian Amazonia: Implications for the Aluminum Industry. *World Development, 77*, 48–65. https://doi.org/10.1016/J.WORLDDEV.2015.08.015
42. Raadal, H. L., Gagnon, L., Modahl, I. S., & Hanssen, O. J. (2011). Life cycle greenhouse gas (GHG) emissions from the generation of wind and hydro power. *Renewable and Sustainable Energy Reviews, 15*(7), 3417–3422. https://doi.org/10.1016/J.RSER.2011.05.001
43. Garg, D., Dashmana, A., Verma, C., Garg, A., & Jain, S. (2023). Deep learning-based analysis of process parameters of biodiesel production using biochar as a catalyst. *Materials Today: Proceedings*. https://doi.org/10.1016/j.matpr.2023.01.356
44. Jain, S., Jain, S., Wolf, I. T., Lee, J., & Tong, Y. W. (2015). A comprehensive review on operating parameters and different pretreatment methodologies for anaerobic digestion of municipal solid waste. *Renewable and Sustainable Energy Reviews, 52*, 142–154. https://doi.org/10.1016/j.rser.2015.07.091
45. Thengane, S. K., Kung, K. S., Gomez-Barea, A., & Ghoniem, A. F. (2022). Advances in biomass torrefaction: Parameters, models, reactors, applications, deployment, and market. *Progress in Energy and Combustion Science, 93*, 101040. https://doi.org/10.1016/j.pecs.2022.101040

46. Fazil Khan, M., Garg, A., Jain, S., Dwivedi, G., & Nath Verma, T. (2020). Optimization of low-temperature transesterification of low FFA blend of sunflower oil and algae oil. *Fuel, 279*. https://doi.org/10.1016/j.fuel.2020.118459
47. Garg, A., & Jain, S. (2020). Process parameter optimization of biodiesel production from algal oil by response surface methodology and artificial neural networks. *Fuel, 277*. https://doi.org/10.1016/j.fuel.2020.118254
48. Jain, S. (2023). Biodiesel production from food waste using insitu transesterification method. *Sustainable Energy Technologies and Assessments, 58*, 103380. https://doi.org/10.1016/j.seta.2023.103380
49. Chen, A., & Liu, Y. (2023). Optimizing sustainable biomass–coal co-firing power plant location problem under ambiguous supply. *Computers & Industrial Engineering, 182*, 109401. https://doi.org/10.1016/j.cie.2023.109401
50. Garg, A., & Jain, S. (2021). *Different types of self-repairing of composite materials: An overview*. https://doi.org/10.1016/B978-0-12-822291-1.00011-7
51. Gesellschaft Energietechnik, Strom und Wärme aus biogenen Festbrennstoffen Fachtagung Salzburg, 15. und 16. Oktober 2008.
52. Adamu, H., et al. (2023). Production processes, techno-economic and policy challenges of bioenergy production from fruit and vegetable wastes. *Renewable and Sustainable Energy Reviews, 186*, 113686. https://doi.org/10.1016/j.rser.2023.113686
53. Garg, A., Chauhan, B. V. S., Vedrantam, A., Jain, S., & Bharti, S. (2022). *Potential and challenges of using biodiesel in a compression ignition engine*. https://doi.org/10.1007/978-981-16-8414-2_9
54. Garg, A., Dwivedi, G., Jain, S., & Behura, A. K. (2021). *Impact of methanol on engine performance and emissions*. https://doi.org/10.1007/978-981-16-1280-0_10
55. Li, M., Fu, Q., Singh, V. P., Liu, D., & Li, J. (2020). Optimization of sustainable bioenergy production considering energy-food-water-land nexus and livestock manure under uncertainty. *Agricultural Systems, 184*, 102900. https://doi.org/10.1016/j.agsy.2020.102900
56. Jain, S. (2023). An assessment of the operation and emission characteristics of a diesel engine powered by a new biofuel prepared using in situ transesterification of a dry spirogyra Algae–Jatropha powder mixture. *Energies (Basel), 16*(3). https://doi.org/10.3390/en16031470
57. Kaleshwarwar, A., & Bahadure, S. (2023). Assessment of the solar energy potential of diverse urban built forms in Nagpur, India. *Sustainable Cities and Society, 96*, 104681. https://doi.org/10.1016/J.SCS.2023.104681
58. Dissanayake, P. D., et al. (2023). Environmental impact of metal halide perovskite solar cells and potential mitigation strategies: a critical review. *Environmental Research, 219*, 115066. https://doi.org/10.1016/J.ENVRES.2022.115066
59. Schoeneberger, C. A., McMillan, C. A., Kurup, P., Akar, S., Margolis, R., & Masanet, E. (2020). Solar for industrial process heat: A review of technologies, analysis approaches, and potential applications in the United States. *Energy, 206*, 118083. https://doi.org/10.1016/J.ENERGY.2020.118083
60. Zhang, Q., Liu, C., & Zheng, S. (2023). Investment and pricing in solar photovoltaic waste recycling with government intervention: A supply chain perspective. *Computers & Industrial Engineering, 177*, 109044. https://doi.org/10.1016/J.CIE.2023.109044
61. Jain, S., Sharma, T., & Gupta, A. K. (2022). End-of-life management of solar PV waste in India: Situation analysis and proposed policy framework. *Renewable and Sustainable Energy Reviews, 153*, 111774. https://doi.org/10.1016/J.RSER.2021.111774
62. Sedighkia, M., & Abdoli, A. (2023). An optimization approach for managing environmental impacts of generating hydropower on fish biodiversity. *Renewable Energy, 218*, 119283. https://doi.org/10.1016/J.RENENE.2023.119283
63. Bildirici, M. E., & Gökmenoğlu, S. M. (2017). Environmental pollution, hydropower energy consumption and economic growth: Evidence from G7 countries. *Renewable and Sustainable Energy Reviews, 75*, 68–85. https://doi.org/10.1016/J.RSER.2016.10.052
64. Mahmud, M. A. P., Farjana, S. H., Lang, C., & Huda, N. (2023). Environmental impacts of hydropower plants. In *Green energy* (pp. 73–102). https://doi.org/10.1016/B978-0-32-385953-0.00010-0

65. Lai, R., Chen, X., & Zhang, L. (2022). Evaluating the impacts of small cascade hydropower from a perspective of stream health that integrates eco-environmental and hydrological values. *Journal of Environmental Management, 305*, 114366. https://doi.org/10.1016/J.JENVMAN.2021.114366
66. Anshelm, J., & Simon, H. (2016). Power production and environmental opinions—Environmentally motivated resistance to wind power in Sweden. *Renewable and Sustainable Energy Reviews, 57*, 1545–1555. https://doi.org/10.1016/J.RSER.2015.12.211
67. Nazir, M. S., Mahdi, A. J., Bilal, M., Sohail, H. M., Ali, N., & Iqbal, H. M. N. (2019). Environmental impact and pollution-related challenges of renewable wind energy paradigm—A review. *Science of The Total Environment, 683*, 436–444. https://doi.org/10.1016/J.SCITOTENV.2019.05.274
68. Loss, S. R., Will, T., & Marra, P. P. (2013). Estimates of bird collision mortality at wind facilities in the contiguous United States. *Biological Conservation, 168*, 201–209. https://doi.org/10.1016/J.BIOCON.2013.10.007
69. Abbasi, T., & Abbasi, S. A. (2010). Biomass energy and the environmental impacts associated with its production and utilization. *Renewable and Sustainable Energy Reviews, 14*(3), 919–937. https://doi.org/10.1016/J.RSER.2009.11.006
70. Hussin, F., Hazani, N. N., Khalil, M., & Aroua, M. K. (2023). Environmental life cycle assessment of biomass conversion using hydrothermal technology: A review. *Fuel Processing Technology, 246*, 107747. https://doi.org/10.1016/J.FUPROC.2023.107747
71. Saravanan, A., et al. (2023). Techno-economic and environmental sustainability prospects on biochemical conversion of agricultural and algal biomass to biofuels. *Journal of Cleaner Production, 414*, 137749. https://doi.org/10.1016/J.JCLEPRO.2023.137749
72. Lee, M., Lin, Y. L., Te Chiueh, P., & Den, W. (2020). Environmental and energy assessment of biomass residues to biochar as fuel: A brief review with recommendations for future bioenergy systems. *Journal of Cleaner Production, 251*, 119714. https://doi.org/10.1016/J.JCLEPRO.2019.119714
73. Entezari, A., Aslani, A., Zahedi, R., & Noorollahi, Y. (2023). Artificial intelligence and machine learning in energy systems: A bibliographic perspective. *Energy Strategy Reviews, 45*, 101017. https://doi.org/10.1016/J.ESR.2022.101017
74. Omitaomu, O. A., & Sayed Shah Danish, M. (2023). AI in energy: Overcoming unforeseen obstacles. *AI 4*(2), 406–425. https://doi.org/10.3390/AI4020022
75. Dellosa, J. T., & Palconit, E. C. (2021). Artificial Intelligence (AI) in renewable energy systems: A condensed review of its applications and techniques. In *21st IEEE International Conference on Environment and Electrical Engineering and 2021 5th IEEE Industrial and Commercial Power System Europe, EEEIC/I and CPS Europe 2021—Proceedings*. https://doi.org/10.1109/EEEIC/ICPSEUROPE51590.2021.9584587
76. Shin, W., Han, J., & Rhee, W. (2021). AI-assistance for predictive maintenance of renewable energy systems. *Energy, 221*, 119775. https://doi.org/10.1016/J.ENERGY.2021.119775
77. Salehi, M., & Co, C. (2023). Revolutionizing renewable energy: AI's role in grid optimization and sustainable future. Accessed: November 15, 2023 [Online]. Available: https://www.oracle.com/industries/utilities/products/opower-energy/
78. Ahmad, T., et al. (2021). Artificial intelligence in sustainable energy industry: Status Quo, challenges and opportunities. *Journal of Cleaner Production, 289*, 125834. https://doi.org/10.1016/J.JCLEPRO.2021.125834
79. Chen, L. et al. (2023). Artificial intelligence-based solutions for climate change: a review. *Environmental Chemistry Letters, 21*(5), 2525–2557. https://doi.org/10.1007/S10311-023-01617-Y
80. The Power to Change Solar and Wind Cost Reduction Potential to 2025. Accessed: November 1, 2023 [Online]. Available: https://www.irena.org/publications/2016/Jun/The-Power-to-Change-Solar-and-Wind-Cost-Reduction-Potential-to-2025
81. Sakiru Folarin Bello. (2024). Ifeoluwa Uchechukwu Wada, Olukayode B Ige, Ernest C Chianumba, and Samod Adetunji Adebayo, "AI-driven predictive maintenance and optimization of renewable energy systems for enhanced operational efficiency and longevity."

International Journal of Science and Research Archive, 13(1), 2823–2837. https://doi.org/10.30574/ijsra.2024.13.1.1992
82. Afridi, Y. S., Ahmad, K., & Hassan, L. (2022). Artificial intelligence based prognostic maintenance of renewable energy systems: A review of techniques, challenges, and future research directions. *International Journal of Energy Research, 46*(15), 21619–21642. https://doi.org/10.1002/ER.7100
83. Surucu, O., Gadsden, S. A., & Yawney, J. (2023). Condition monitoring using machine learning: A review of theory, applications, and recent advances. *Expert Systems with Applications, 221*, 119738. https://doi.org/10.1016/j.eswa.2023.119738
84. Ahmed, I. A., & Asamoah, P. B. (2024). AI-driven predictive maintenance for energy infrastructure. International Journal of Research and Scientific Innovation, XI, 507–528. https://doi.org/10.51244/IJRSI.2024.1109048
85. Dereci, U., & Tuzkaya, G. (2024). An explainable artificial intelligence model for predictive maintenance and spare parts optimization. *Supply Chain Analytics, 8*, 100078. https://doi.org/10.1016/j.sca.2024.100078
86. Gama, J., Ribeiro, R. P., Mastelini, S., Davari, N., & Veloso, B. (2024). From fault detection to anomaly explanation: A case study on predictive maintenance. *Journal of Web Semantics, 81*, 100821. https://doi.org/10.1016/j.websem.2024.100821
87. Hulwan, D. B., et al. (2024). AI-based fault detection and predictive maintenance in wind power conversion systems. In *E3S Web of Conferences* (vol. 591, p. 02003). https://doi.org/10.1051/e3sconf/202459102003
88. Wilson, H. J., & Daugherty, P. R. *Collaborative intelligence: Humans and AI are joining forces*.
89. Ahmad, T., et al. (2022). Energetics systems and artificial intelligence: Applications of industry 4.0. *Energy Reports, 8*, 334–361. https://doi.org/10.1016/J.EGYR.2021.11.256
90. Improve operational decisions with longer range forecasts—Met Office. Accessed: November 9, 2023 [Online]. Available: https://www.metoffice.gov.uk/services/insights/improve-operational-decisions
91. Ghalehkhondabi, I., Ardjmand, E., Weckman, G. R., & Young, W. A. (2017). An overview of energy demand forecasting methods published in 2005–2015. *Energy Systems, 8*(2), 411–447. https://doi.org/10.1007/S12667-016-0203-Y/TABLES/8
92. Paletta, Q., et al. (2023). Advances in solar forecasting: Computer vision with deep learning. *Advances in Applied Energy, 11*, 100150. https://doi.org/10.1016/J.ADAPEN.2023.100150
93. Bi, K., Xie, L., Zhang, H., Chen, X., Gu, X., & Tian, Q. (2023). Accurate medium-range global weather forecasting with 3D neural networks. *Nature 619*(7970), 533–538. https://doi.org/10.1038/s41586-023-06185-3
94. Benti, N. E., Chaka, M. D., & Semie, A. G. (2023). Forecasting renewable energy generation with machine learning and deep learning: Current advances and future prospects. *Sustainability, 15*(9), 7087. https://doi.org/10.3390/SU15097087
95. Voyant, C., et al. (2017). Machine learning methods for solar radiation forecasting: A review. *Renewable Energy, 105*, 569–582. https://doi.org/10.1016/J.RENENE.2016.12.095
96. Srinivasan, D. (2008). Energy demand prediction using GMDH networks. *Neurocomputing, 72*(1–3), 625–629. https://doi.org/10.1016/J.NEUCOM.2008.08.006
97. Majeed Butt, O., Zulqarnain, M., & Majeed Butt, T. (2021). Recent advancement in smart grid technology: Future prospects in the electrical power network. *Ain Shams Engineering Journal, 12*(1), 687–695. https://doi.org/10.1016/J.ASEJ.2020.05.004
98. Bari, A., Jiang, J., Saad, W., & Jaekel, A. (2014). Challenges in the smart grid applications: An overview. https://doi.org/10.1155/2014/974682
99. Alsuwian, T., Shahid Butt, A., & Amin, A. A. (2022). Smart grid cyber security enhancement: Challenges and solutions—A review. *Sustainability, 14*, 14226. https://doi.org/10.3390/SU142114226
100. Nazmul Hasan, M., Toma, R. N., Al Nahid, A., Manjurul Islam, M. M., & Kim, J. M. (2019). Electricity theft detection in smart grid systems: A CNN-LSTM based approach. *Energies, 12*(17), 3310. https://doi.org/10.3390/EN12173310

101. Wang, X., Wang, H., Bhandari, B., & Cheng, L. (2023). AI-empowered methods for smart energy consumption: A review of load forecasting, anomaly detection and demand response. *International Journal of Precision Engineering and Manufacturing-Green Technology, 2023*, 1–31. https://doi.org/10.1007/S40684-023-00537-0
102. Dorji, S., Stonier, A. A., Peter, G., Kuppusamy, R., & Teekaraman, Y. (2023). An extensive critique on smart grid technologies: recent advancements, key challenges, and future directions. *Technologies, 11*(3), 81. https://doi.org/10.3390/TECHNOLOGIES11030081
103. Smart grids—IEA. Accessed: November 15, 2023 [Online]. Available: https://www.iea.org/energy-system/electricity/smart-grids#tracking
104. Raza, M. A., Aman, M. M., Abro, A. G., Tunio, M. A., Khatri, K. L., & Shahid, M. (2022). Challenges and potentials of implementing a smart grid for Pakistan's electric network. *Energy Strategy Reviews, 43*, 100941. https://doi.org/10.1016/J.ESR.2022.100941
105. Şalvarlı, D.-D. M. A. D. (2023). Some views on renewable energy marketing due to climate change impacts. *World Journal of Multidisciplinary Research, 6*(1), 161–180. https://doi.org/10.58853/DUMAD.1233721
106. Bahlawan, H., et al. (2021). Optimal design and energy management of a renewable energy plant with seasonal energy storage. In *E3S Web of Conferences, 238*, 02002. https://doi.org/10.1051/E3SCONF/202123802002
107. Abdullah, S., Elshaikh, M., & Mahmood, M. B. (2021). Comparative study of energy storage systems (ESSs) you may also like food code breaker (FCB)-developing the multi-lingual food code translation android application using multi-options code reader system. https://doi.org/10.1088/1742-6596/1962/1/012035
108. IEEE Xplore Full-Text PDF. Accessed: November 13, 2023 [Online]. Available: https://ieeexplore.ieee.org/stamp/stamp.jsp?arnumber=9076660
109. The Future of Energy Storage An Interdisciplinary MIT Study.
110. Li, Y., et al. (2019). Data-driven health estimation and lifetime prediction of lithium-ion batteries: A review. *Renewable and Sustainable Energy Reviews, 113*, 109254. https://doi.org/10.1016/J.RSER.2019.109254
111. Brynjolfsson, E., & Mitchell, T. (2017). What can machine learning do? Workforce implications: Profound change is coming, but roles for humans remain. *Science, 358*(6370), 1530–1534. https://doi.org/10.1126/SCIENCE.AAP8062/SUPPL_FILE/AAP8062-BRYNJOLFSSON-SM.PDF
112. Franki, V., Majnarić, D., & Višković, A. (2023). A comprehensive review of Artificial Intelligence (AI) companies in the power sector. *Energies, 16*(3), 1077. https://doi.org/10.3390/EN16031077
113. Artificial intelligence for the power grid of the future | umlaut. Accessed: December 03, 2023 [Online]. Available: https://www.umlaut.com/en/stories/artificial-intelligence-for-the-power-grid-of-the-future
114. Krause, T., Ernst, R., Klaer, B., Hacker, I., & Henze, M. (2021). Cybersecurity in power grids: challenges and opportunities. *Sensors (Basel), 21*(18). https://doi.org/10.3390/S21186225
115. Szczepaniuk, H., & Szczepaniuk, E. K. (2022). Applications of artificial intelligence algorithms in the energy sector. *Energies, 16*, 347. https://doi.org/10.3390/EN16010347
116. Cambini, C., Meletiou, A., Bompard, E., & Masera, M. (2016). Market and regulatory factors influencing smart-grid investment in Europe: Evidence from pilot projects and implications for reform. *Util Policy, 40*, 36–47. https://doi.org/10.1016/J.JUP.2016.03.003
117. Vinuesa, R., et al. (2020). The role of artificial intelligence in achieving the sustainable development goals. *Nature Communications 11*(1), 1–10. https://doi.org/10.1038/s41467-019-14108-y
118. The International Renewable Energy Agency. (2019). *Artificial intelligence and big data innovation landscape brief*. Accessed: December 3, 2023 [Online]. Available: www.irena.org
119. Goralski, M. A., & Tan, T. K. (2020). Artificial intelligence and sustainable development. *The International Journal of Management Education, 18*(1), 100330. https://doi.org/10.1016/J.IJME.2019.100330

Chapter 4
Amalgamation of Artificial Intelligence and Climate Change

Balendra V. S. Chauhan, Ajitanshu Vedrtnam, Kevin P. Wyche, and Sneha Verma

Abstract This chapter explores the combination of artificial intelligence (AI) and climate modeling, highlighting how AI can improve climate predictions. Traditional climate models, which are based on physics, have been important for our climate understanding, but they struggle to fully capture the complexities of the Earth's climate system, especially at detailed local scales and in predicting extreme weather events. AI, by using its strengths in recognizing patterns and analyzing data, offers a complementary approach to address these limitations. The sources describe various AI techniques used in climate prediction, including supervised learning methods for forecasting based on historical data; unsupervised learning algorithms for identifying distinct climate patterns; deep learning architectures for high-resolution spatial and temporal predictions; and hybrid models that combine AI with traditional physics-based models to improve accuracy and reliability. These AI-driven approaches enhance climate downscaling, El Niño event prediction, sea-level rise projections, and extreme weather forecasting. However, integrating AI into climate science comes with challenges: needing large, high-quality datasets for training; requiring significant computing power; needing to understand how AI models make predictions; and ethical considerations to avoid biases. The chapter emphasizes future research focusing on developing hybrid models that blend AI with physical principles; enhancing model interpretability; and addressing data gaps. By

B. V. S. Chauhan · K. P. Wyche (✉)
Centre for Earth Observation Science, University of Brighton, Brighton, UK
e-mail: kevinpwyche@googlemail.com

B. V. S. Chauhan
e-mail: b.chauhan@brighton.ac.uk

A. Vedrtnam
Institute of Construction and Architecture, Slovak Academy of Science, Bratislava, Slovakia

Department of Mechanical Engineering, Invertis University, Bareilly, India

A. Vedrtnam
e-mail: ajitanshu.m@invertis.org

B. V. S. Chauhan · S. Verma
School of Engineering, Newcastle University, Newcastle Upon Tyne, United Kingdom
e-mail: Sneha.verma@newcastle.ac.uk

A. Vedrtnam et al., *Prospects of Artificial Intelligence in the Environment*,
https://doi.org/10.1007/978-981-96-6863-2_4

tackling these challenges, AI-driven climate prediction can significantly improve our understanding of climate dynamics and lead to effective policy responses to climate change.

Keywords Artificial Intelligence (AI) · Environment · Climate science · Environmental monitoring · Climate modeling · Prediction · El Niño event · Ecosystem management · Natural disaster management · Extreme weather forecasting · Waste management · Conservation · Data gaps

1 Introduction

Climate modeling and prediction have become indispensable tools in the global fight against climate change, forming the scientific backbone for understanding complex climate dynamics, forecasting future scenarios, and guiding critical policy decisions on mitigation and adaptation strategies. As climate change accelerates, manifesting in an alarming increase in extreme weather events, rising sea levels, melting ice caps, and shifting ecosystems, the need for precise and reliable climate predictions has never been more urgent. These models not only help us understand the causes and consequences of climate variability but also enable policymakers to implement strategies that can mitigate the worst impacts of global warming. The Intergovernmental Panel on Climate Change (IPCC) reports that global surface temperatures have already risen by approximately 1.1 °C since the pre-industrial era, a trend that shows no sign of slowing. Without significant global efforts to reduce greenhouse gas emissions, projections suggest that we could exceed the critical 1.5 °C threshold as early as 2030, intensifying the risk of irreversible changes to the Earth's systems. Understanding these projections, especially at a regional level, is crucial for disaster preparedness, infrastructure planning, food security, and water resource management [1].

Traditionally climate models have been grounded in physics-based approaches, rooted in the laws of thermodynamics, fluid dynamics, and radiative transfer to simulate the interactions among key components of the Earth's system such as the atmosphere, oceans, land surface, and ice. These models have evolved in complexity, from simple Energy Balance Models (EBMs) [2] that focus primarily on global temperature changes to General Circulation Models (GCMs) [3] and Earth System Models (ESMs) [4] that account for detailed processes like ocean currents, atmospheric chemistry, and biogeochemical cycles. GCMs and ESMs have been invaluable in identifying broad trends in climate behavior under various emissions scenarios, such as the projected increase in the frequency and intensity of extreme weather events like hurricanes, droughts, and floods [5–7]. However, while these traditional models have offered crucial insights, they face limitations in capturing the complex, nonlinear interactions that characterize the climate system particularly at finer spatial scales. Their coarse resolution often means that local climate phenomena, such as urban heat islands, regional monsoons, or the impacts of localized land-use changes, are not accurately represented. Moreover, predicting extreme weather events, which involve highly complex atmospheric dynamics, remains a significant challenge for traditional models.

In recent years, the emergence of AI and ML has revolutionized climate modeling. By leveraging vast amounts of observational data, AI-driven approaches complement and enhance traditional physics-based models, introducing a paradigm shift in how we process and interpret climate information. AI allows for the identification of intricate patterns within large and complex datasets, which can improve the accuracy of predictions, particularly in areas where traditional models have struggled. For instance, deep learning models like Convolutional Neural Networks (CNNs) and Recurrent Neural Networks (RNNs) have been particularly effective in downscaling low-resolution outputs from GCMs to produce highly detailed regional forecasts. This is crucial for local governments and communities to effectively prepare for climate impacts such as coastal flooding, agriculture disruption, and energy demands [8–11]. Additionally, AI techniques such as unsupervised learning can classify climate patterns like the El Niño-Southern Oscillation (ENSO), while generative models like Generative Adversarial Networks (GANs) are used to create synthetic climate data for scenario analysis. This integration of AI not only helps in improving spatial and temporal resolution but also enhances our understanding of the probability and severity of future climate events [12, 13]. Nonetheless, the incorporation of AI into climate science introduces its own set of challenges. AI models are highly dependent on the quality, quantity, and diversity of the data they are trained on. Climate data often come from heterogeneous sources, including satellite observations, weather stations, ocean buoys, and reanalysis products, each with its own biases and uncertainties. Data sparsity in developing regions, where weather and climate monitoring infrastructure may be lacking, remains a significant hurdle. Furthermore, the computational demands of training AI models on vast, high-resolution datasets require advanced high-performance computing (HPC) infrastructure, which is not always accessible to all researchers [14, 15].

Moreover, while AI-driven models offer significant improvements in prediction accuracy, they are often criticized for being "black boxes" lacking the transparency and interpretability required for high-stakes decision-making [16, 17]. Ensuring that AI predictions align with well-understood physical principles is crucial to maintaining scientific rigor and trust in these models. To that end, hybrid models that combine AI's predictive power with the interpretability of traditional physics-based methods are emerging as a promising solution. These physics-informed AI models ensure that key physical laws, such as conservation of energy or mass, are embedded into AI algorithms, enhancing the credibility and robustness of the results. This chapter explores the evolution of climate modeling through the integration of AI and ML techniques. We will highlight the distinctions between traditional models and AI-enhanced approaches, delve into the specific types of models and AI techniques used in climate prediction, and discuss the advantages and limitations of each. Additionally, we will examine the significant computational, data integration, and validation challenges that must be addressed to fully harness the potential of AI in advancing climate science. By providing a comprehensive overview of these developments, this chapter aims to underscore the transformative role of AI in improving our understanding of climate dynamics and enhancing our ability to predict and respond to the increasingly complex and urgent challenges posed by climate change.

2 Overview of Climate Modeling

2.1 Introduction to Climate Modelling

Climate modeling refers to the use of quantitative tools and techniques to simulate the behavior and interactions of the Earth's climate system [18]. These models are grounded in fundamental physical laws, such as thermodynamics, which governs energy transfers, fluid dynamics, which describes the movement of the atmosphere and oceans, and radiative transfer, which focuses on the exchange of radiation between the Earth's surface and the atmosphere. By representing the interactions among the key components of the climate system atmosphere, oceans, land surface, ice, and biogeochemical cycles these models simulate historical climate behavior and project future climate scenarios. Climate models are used to understand how external and internal forcing agents influence climate variability and long-term trends. Forcing scenarios refer to specific inputs that drive changes in the climate system. These scenarios allow scientists to "stress test" climate models by simulating how the Earth's climate might respond under different conditions, such as varying levels of greenhouse gas (GHG) emissions, aerosol concentrations, volcanic activity, or land-use changes like deforestation. One widely used set of forcing scenarios is the Representative Concentration Pathways (RCPs), developed for the Intergovernmental Panel on Climate Change (IPCC) [19]. RCPs outline different pathways based on future greenhouse gas concentrations and radiative forcing levels, ranging from a low-emission scenario (RCP 2.6) to a high-emission scenario (RCP 8.5) [20]. These scenarios help researchers simulate how future climate might behave under varying levels of human activity, enabling policymakers to explore the potential outcomes of different mitigation and adaptation strategies.

For example, under the RCP 8.5 scenario, which represents a "business-as-usual" approach with high emissions, climate models project substantial warming, more frequent extreme weather events, and significant sea-level rise [21]. In contrast, under RCP 2.6, which assumes aggressive emission reductions, the models project much lower levels of warming and a reduction in severe climate impacts [22]. These scenarios enable scientists to model different futures, offering insights into the range of possible outcomes based on current policy and behavior choices. Climate modeling is an iterative process [23]. As new data, computational techniques, and scientific insights become available, models are continuously refined and updated. The development of climate models involves running simulations based on historical data to ensure that models accurately represent past climate behavior. This process of hindcasting, or testing models against past observed climate conditions, helps modelers fine-tune their algorithms and correct biases. The iterative nature of this process allows for a continuous feedback loop where observational data informs model improvement, and improved models yield more accurate predictions.

This iterative refinement also benefits from advancements in high-performance computing (HPC), which allow models to simulate climate at higher resolutions and with greater complexity [24]. As computational power grows, scientists can integrate more detailed representations of physical processes, such as cloud formation, ocean circulation, and biogeochemical cycles, resulting in more robust and reliable climate projections. One of the most critical aspects of climate modeling is the use of observational data to both calibrate and validate models. Observational data comes from a wide array of sources, including satellites, weather stations, ocean buoys, ice core samples, and paleoclimate proxies. These data are used to ensure that the models accurately representing real-world conditions. For example, models are often tested against satellite observations of sea surface temperatures, atmospheric carbon dioxide concentrations, or the extent of polar ice to verify their accuracy. Moreover, observational data is essential for detecting and correcting biases in climate models [25]. These biases arise because even the most sophisticated models simplify certain processes or omit small-scale dynamics. By comparing model outputs with observed data, scientists can make necessary adjustments to improve the fidelity of their projections. In addition to calibration, observational data is crucial for model validation. Validation involves assessing how well a climate model's projections align with independent datasets that were not used during the model's development. This helps ensure that the model's predictions are not only accurate for the conditions it was trained on but also for unseen data. Model validation is often done using techniques such as cross-validation or by comparing models with observational datasets from different regions, periods, or specific events like El Niño or major volcanic eruptions [26].

As new observational data becomes available and our understanding of the climate system improves, models are continuously updated and refined. This process ensures that climate models remain relevant and incorporate the latest scientific findings. For instance, advances in satellite technology have significantly increased the quantity and quality of climate data, allowing for better resolution and coverage of key variables like aerosol concentrations, land cover, and sea-level rise. Similarly, improved algorithms and numerical techniques allow models to simulate more complex interactions within the climate system. For example, newer models are better able to represent cloud microphysics, ocean–atmosphere coupling, and carbon cycle feedback, all of which are critical for understanding long-term climate dynamics [27]. Additionally, the development of hybrid models that integrate ML and physics-based approaches has further enhanced climate prediction capabilities, enabling models to capture nonlinear processes and extreme events that are challenging to represent using traditional methods alone [28]. So, climate models are dynamic, quantitative representations of the Earth's climate system [29]. They are continuously improved through iterative cycles of testing and validation against real-world data. By incorporating various forcing scenarios and observational data, climate models provide vital insights into both historical climate behavior and future climate projections. As climate science advances and computational tools become more powerful, climate models will continue to play a critical role in informing the global response to climate change.

2.2 *Types of Climate Models*

Climate models have evolved significantly over time, starting with simple energy balance approaches and advancing to complex Earth system models. Each type of model serves different purposes, ranging from understanding basic climate mechanisms to informing international climate policy decisions. The evolution of climate modeling reflects the increasing complexity of our understanding of the climate system and the need for models to capture diverse processes at varying spatial and temporal scales.

Energy Balance Models (EBMs)

Energy Balance Models (EBMs) are some of the simplest climate models, designed to simulate the Earth's temperature by solving conservation of energy equations [30]. They operate on the principle that the energy absorbed by the Earth (from the sun) must balance the energy radiated back into space. These models typically treat the Earth as a single "box" or as a series of latitude zones, calculating the average temperature of the planet without accounting for finer spatial details or many physical processes. EBMs are particularly useful in early-stage research and conceptual studies, as they provide a first-order understanding of climate sensitivity and the response of global temperatures to changes in radiative forcing, such as increased greenhouse gas (GHG) concentrations [31]. However, their simplicity means they cannot capture the full complexity of the Earth system. For instance, they ignore critical interactions between the atmosphere, oceans, ice sheets, and land surface, and they do not include processes such as cloud formation, ocean currents, or feedback mechanisms that are crucial in more detailed models. While EBMs offer valuable insights, especially in understanding the energy balance of the planet, their limited ability to represent spatial variability makes them unsuitable for regional climate predictions or long-term projections [32]. Historically, EBMs laid the groundwork for more sophisticated models, serving as a starting point for understanding global temperature dynamics. However, as climate science advanced and computing power increased, models expanded beyond simple energy balances to include the complex processes captured in General Circulation Models (GCMs) and Earth System Models (ESMs).

General Circulation Models (GCMs)

General Circulation Models (GCMs) are much more sophisticated than EBMs [33]. They simulate the circulation of the Earth's atmosphere and oceans by solving the fundamental equations of fluid dynamics, thermodynamics, and radiative transfer over a three-dimensional grid that spans the globe. GCMs divide the Earth's atmosphere and oceans into a network of grid cells, where each cell represents a small part of the Earth's surface or atmosphere. Within these grid cells, the model solves equations to track the transfer of heat, moisture, momentum, and other key variables between adjacent cells.

The spatial resolution of GCMs is a critical factor in their accuracy [34]. A higher resolution (with smaller grid cells) allows for more detailed representations of the climate system, capturing phenomena like cloud formation or ocean eddies more accurately. However, increasing resolution also requires vastly more computational resources. The coarser grid cells in lower-resolution models can miss local climate features, such as regional temperature variations or small-scale atmospheric patterns, though they are still excellent at simulating broad global trends like global temperature rise, polar ice melt, and shifts in large-scale circulation patterns like the jet stream. GCMs are widely used for international climate assessments, including those by the Intergovernmental Panel on Climate Change (IPCC) [35]. For example, the IPCC's Coupled Model Intercomparison Project (CMIP), currently in its sixth phase (CMIP6), includes many GCMs developed by research institutions worldwide [36]. These models are used to simulate climate responses under different emission scenarios (e.g., RCPs or Shared Socioeconomic Pathways (SSPs)) and provide the scientific basis for global climate projections [37]. GCMs are highly computationally intensive, often requiring the use of supercomputers to run simulations. The need for such computational power is due to the vast number of grid cells and the complexity of the calculations involved. In recent years, GCMs have incorporated increasingly detailed sub-grid parameterizations to better represent processes like cloud dynamics, sea ice behavior, and vegetation.

Regional Climate Models (RCMs)

While GCMs are excellent at simulating global-scale climate processes, their coarse resolution limits their ability to capture regional climate dynamics [38]. To address this limitation, Regional Climate Models (RCMs) have been developed. RCMs are high-resolution models that focus on smaller geographic areas, often "nested" within GCMs to provide more detailed simulations for specific regions [39]. By zooming in on regions of interest, such as coastal zones, mountain ranges, or urban areas, RCMs can resolve local climate phenomena that GCMs may miss, such as urban heat islands, monsoon patterns, or localized precipitation events. RCMs are especially valuable for local adaptation planning and disaster preparedness in regions that are particularly vulnerable to climate impacts, such as South Asia, Sub-Saharan Africa, or the Arctic. For instance, RCMs can provide highly localized projections of temperature and precipitation patterns, which are critical for managing water resources, agricultural planning, and infrastructure development in response to anticipated climate changes. RCMs work by taking the boundary conditions provided by GCMs (such as temperature, pressure, and wind patterns) and refining these inputs to simulate local conditions at a much higher resolution [40]. This "downscaling" process allows for more accurate representations of regional climates, making RCMs essential tools for city planners, policy makers, and conservationists who need to understand how climate change will affect specific regions [41].

Earth System Models (ESMs)

Earth System Models (ESMs) represent the most advanced type of climate model, expanding upon the physical processes captured by GCMs to include biogeochemical cycles and interactions between the climate system and biological, chemical, and geological processes [42]. Figure 1 represents the basic earth system model. In addition to simulating the atmosphere, oceans, and land surface, ESMs incorporate processes like the carbon cycle, nitrogen cycle, vegetation dynamics, and ocean biology [43, 44]. This allows them to simulate feedback loops, such as how changes in carbon emissions affect the growth of forests or how ocean acidification impacts marine ecosystems.

One of the key strengths of ESMs is their ability to model carbon-climate feedback, which are critical for long-term climate projections [45]. For example, as global temperatures rise, permafrost in the Arctic may thaw, releasing methane (a potent greenhouse gas) into the atmosphere, which could further accelerate warming. ESMs are capable of simulating such complex interactions and are therefore essential for understanding the potential tipping points in the climate system points at which small changes could lead to dramatic, irreversible shifts in climate. ESMs are also used to

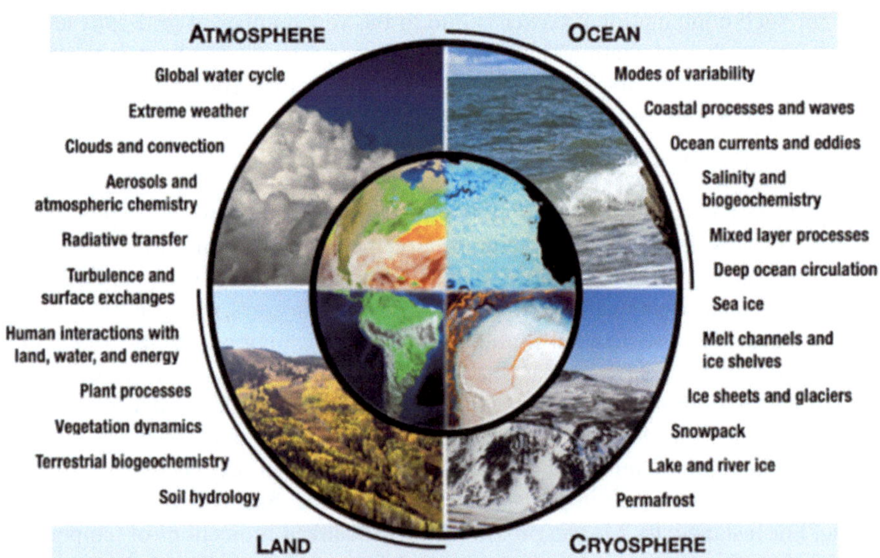

Fig. 1 Illustration of basic earth system model

predict land-use changes, such as deforestation or reforestation, and their impact on climate. By modeling the interactions between human activities and natural systems, ESMs provide valuable insights into how different policy decisions such as reducing deforestation or increasing renewable energy adoption can influence the climate [46]. Examples of widely used ESMs include the Community Earth System Model (CESM) developed by the National Center for Atmospheric Research (NCAR) and the UK Earth System Model (UKESM) [47, 48]. These models are integral to the work of the IPCC and other international climate research efforts, providing the most comprehensive and detailed simulations of future climate scenarios.

The Importance of Spatial Resolution and Computational Power

Across all types of climate models, spatial resolution is a critical factor in determining the level of detail and accuracy of climate projections. While simpler models like EBMs provide broad global insights, more complex models like GCMs and ESMs offer much finer spatial resolution, allowing for more detailed simulations. However, higher resolution comes with increased computational demands. Running high-resolution models requires powerful supercomputers, as well as advanced numerical techniques to efficiently solve the complex equations governing the climate system. Increasing the resolution of models improves the accuracy of simulations, particularly for regional climate features and extreme weather events like hurricanes, floods, and droughts. This has become especially important as policymakers and planners seek to address the localized impacts of climate change. Machine learning and artificial intelligence are also being explored to enhance the efficiency and resolution of climate models, opening new frontiers for climate downscaling and prediction [49–51]. The climate models have evolved from the simple energy balance approaches of EBMs to the complex Earth system simulations of ESMs. Each type of model serves a specific purpose, from understanding basic climate dynamics to predicting long-term interactions between human activities and the environment. As computational power grows and our understanding of the climate system deepens, these models will continue to provide critical insights into the future of our planet [52].

2.3 *Components of Climate Models*

Climate models consist of various components representing the different parts of the Earth's system. Each component plays a critical role in simulating climate variability and interacting with other components through feedback loops and nonlinear processes. The main components of climate models atmosphere, oceans, land surface, cryosphere, and carbon cycle are interconnected, and changes in one part of the system can trigger significant responses in others [53]. Understanding these interactions is key to improving the accuracy of climate predictions and identifying potential tipping points in the climate system.

Atmospheric Component

The atmosphere is one of the most critical components of climate models, as it governs the Earth's weather and climate patterns. It involves simulating variables such as air temperature, pressure, humidity, wind dynamics, and cloud formation [54]. The atmosphere is the medium through which radiative forcing the balance between incoming solar radiation and outgoing terrestrial radiation occurs, which directly impacts the Earth's surface temperature. One of the most significant challenges in atmospheric modeling is accurately representing cloud dynamics [55]. Clouds play a dual role in the Earth's climate system: they can cool the planet by reflecting incoming sunlight (increasing the Earth's albedo), and they can warm it by trapping outgoing longwave radiation (acting as a greenhouse gas). This duality makes clouds a major source of uncertainty in climate models. Depending on their type, altitude, and coverage, clouds can either amplify or mitigate warming, and this variability is difficult to simulate accurately. Aerosols also play a crucial role in the atmospheric component. Aerosols are tiny particles suspended in the atmosphere that can influence cloud formation and radiative forcing [56–58]. They often act as cloud condensation nuclei, affecting cloud properties and their radiative impact. Aerosols such as sulfates reflect sunlight and have a cooling effect, while others, like black carbon, absorb heat and contribute to warming [59]. The interactions between aerosols and clouds are complex and represent another source of uncertainty in climate predictions. The representation of radiative forcing in climate models also involves accounting for greenhouse gas (GHG) concentrations, aerosol loads, and land-use changes, all of which modify the balance of incoming and outgoing energy. Climate models must accurately simulate the atmospheric circulation patterns (such as Hadley cells and jet streams) that distribute heat around the globe, influencing weather and climate on both regional and global scales.

Oceanic Component

The oceans are another critical component of climate models, playing a major role in regulating the Earth's climate by absorbing and redistributing heat [60]. Oceans are the largest reservoirs of heat and carbon, making ocean heat uptake a key factor in modulating the pace of global warming [61]. The oceans absorb about 90% of the excess heat generated by anthropogenic greenhouse gas emissions, effectively buffering the rate of surface temperature rise [62]. However, this heat does not remain uniformly distributed, and its gradual release back into the atmosphere can significantly influence long-term climate patterns. One of the most important processes represented in oceanic models is thermohaline circulation, which is driven by differences in water density caused by temperature (thermal) and salinity (haline) variations [63, 64]. A critical part of this circulation is the Atlantic Meridional Overturning Circulation (AMOC), which transports warm water from the tropics to the North Atlantic and cold water back to the tropics [65]. AMOC plays a pivotal role in regulating the climate of the North Atlantic region and beyond. If disrupted, it could lead to significant changes in regional climates, such as the potential for more severe winters in Europe and shifts in tropical rain belts. Ocean circulation patterns, such as El Niño-Southern Oscillation (ENSO), also have profound global climate impacts, influencing weather patterns, precipitation, and temperature variability [66, 67]. ENSO is characterized by periodic warming (El Niño) or cooling (La Niña) of the central and eastern Pacific Ocean, leading to disruptions in weather systems across the globe [68]. Accurate representation of these oceanic patterns in climate models is essential for understanding both short-term climate variability and long-term trends. The oceanic carbon cycle is another key element, as the oceans act as a major carbon sink, absorbing significant amounts of atmospheric CO_2. However, as the oceans warm, their capacity to absorb CO_2 decreases, leading to a potential positive feedback loop where more CO_2 remains in the atmosphere, accelerating global warming [69].

Land Surface and Cryosphere

The land surface and cryosphere (the frozen parts of the Earth, including glaciers, ice sheets, sea ice, and permafrost) play crucial roles in influencing the Earth's climate through their interactions with the atmosphere and oceans [70]. One of the primary processes that climate models account for in this component is albedo the reflectivity of the Earth's surface. Snow and ice-covered regions have high albedo, reflecting most incoming solar radiation, while darker surfaces (such as forests or oceans) absorb more heat [71]. As the planet warms and ice melts, albedo decreases, causing more solar radiation to be absorbed by the surface, which accelerates warming [72]. This is known as the ice-albedo feedback, a critical positive feedback loop that exacerbates climate change, particularly in polar regions. The cryosphere also includes processes such as permafrost thaw, which can release trapped methane (a potent greenhouse gas) into the atmosphere, potentially leading to further warming [73]. These processes are important to capture in models, particularly for long-term climate projections. The loss of ice in the Arctic and Antarctic is one of the most visible impacts of climate change and is projected to have significant global consequences, including sea-level rise and changes in ocean circulation [74]. The land surface component also includes vegetation and soil moisture, both of which play roles in regulating climate through processes like evapotranspiration and carbon sequestration. Vegetation feedback are significant; for instance, deforestation reduces the amount of carbon absorbed by trees, contributing to higher atmospheric CO_2 levels. Conversely, reforestation and the growth of new vegetation can act as a carbon sink, absorbing CO_2 from the atmosphere and mitigating climate change [75]. The vegetation-climate feedback is another key area of uncertainty, particularly in the context of human-induced land-use changes. The cryosphere and land surface components are also critical for predicting extreme weather events, such as floods, droughts, and heatwaves, all of which have become more frequent and severe in recent decades [76].

Carbon Cycle

The carbon cycle is a fundamental part of climate models, as it simulates the exchange of carbon between the atmosphere, oceans, and land [77]. This cycle is crucial for understanding how much of the carbon dioxide (CO_2) emitted by human activities is absorbed by natural sinks (oceans and vegetation) and how much remains in the atmosphere, driving climate change [78]. ESMs attempt to model carbon-climate feedback, which are processes that can either amplify or dampen climate change. For instance, as global temperatures rise, permafrost (frozen soil) in the Arctic and sub-Arctic regions may begin to thaw, releasing large amounts of methane (CH_4) and CO_2 that had been stored for millennia. Methane is particularly concerning because it has 28–34 times the global warming potential of CO_2 over a 100-year period [79]. This release of greenhouse gases from thawing permafrost could create a feedback loop where additional warming triggers further methane and CO_2 emissions, leading to runaway climate change. Another important feedback loop involves the ocean's ability to absorb CO_2. As oceans warm, their ability to absorb CO_2 diminishes, leaving more CO_2 in the atmosphere, which further accelerates warming. Models need to accurately represent how carbon sinks such as forests, soils, and oceans will respond to a warming climate and how human activities, such as deforestation, will alter the carbon balance. ESMs also simulate interactions between the carbon cycle and ocean acidification, a process caused by the uptake of CO_2 by the oceans [80]. As more CO_2 dissolves in seawater, it lowers the pH of the water, making it more acidic. This has significant impacts on marine ecosystems, particularly coral reefs and shellfish, and can alter the ocean's capacity to act as a carbon sink.

2.4 Challenges in Climate Modeling

Despite significant advancements in climate modeling, several critical challenges remain that impact the accuracy and reliability of model predictions. These challenges stem from the inherent complexity of the climate system, limitations in computational power, and uncertainties related to feedback mechanisms, spatial and temporal resolution, and long-term projections. Addressing these challenges is crucial for improving climate models and providing more accurate projections of future climate conditions.

Uncertainty in Climate Sensitivity

One of the central challenges in climate modeling is the uncertainty in climate sensitivity, the degree to which the Earth's temperature responds to a given increase in greenhouse gas (GHG) concentrations, typically measured as the temperature change resulting from a doubling of atmospheric CO_2 levels. Current estimates of climate sensitivity range between 1.5 and 4.5 °C, reflecting a substantial degree of uncertainty [81]. This wide range is primarily due to difficulties in accurately modeling cloud formation, aerosol interactions, and feedback mechanisms, which can either amplify or mitigate the effects of increased GHGs. Cloud feedbacks are a major source of this uncertainty. As mentioned in the atmospheric component of climate models, clouds can either cool or warm the planet depending on their type, altitude, and location. Low clouds tend to reflect sunlight, providing a cooling effect, while high clouds trap heat, contributing to warming. The net effect of clouds on climate sensitivity remains a significant unknown. Additionally, aerosols tiny particles in the atmosphere complicate the picture by influencing cloud properties and interacting with sunlight and radiation. Some aerosols, like sulfates, reflect sunlight and cool the Earth, while others, like black carbon, absorb heat and warm the atmosphere [59, 82]. These uncertainties in cloud-aerosol interactions and feedback loops directly impact predictions of future warming. As climate models differ in how they treat these processes, the resulting projections can vary widely, contributing to uncertainty in future temperature increases and related impacts, such as sea-level rise and extreme weather events.

Temporal and Spatial Resolution

Another major challenge in climate modeling is the trade-off between spatial and temporal resolution [83]. Climate models operate on a grid system, where the Earth's surface and atmosphere are divided into grid cells [84]. Within each cell, climate variables such as temperature, pressure, and humidity are calculated. Higher spatial resolution means smaller grid cells and a more detailed representation of the Earth's surface, but this also requires significantly more computational power. Global Climate Models (GCMs), which simulate climate processes on a global scale, typically have coarser spatial resolution, with grid cells covering areas up to 100–250 km wide [85]. This means that while GCMs can accurately model large-scale phenomena, such as global temperature rise or ocean circulation patterns, they often miss local climate phenomena, such as urban heat islands, regional precipitation patterns, or the influence of small-scale topography. This coarse resolution is particularly problematic when trying to model extreme weather events, like hurricanes, tornadoes, or heatwaves, which occur at much smaller scales.

On the other hand, Regional Climate Models (RCMs) are designed to address this issue by providing higher-resolution simulations for specific geographic regions, with grid cells as small as 10 km [86]. However, RCMs typically rely on boundary conditions from GCMs, which means their accuracy is still tied to the broader, coarser GCMs. The challenge is how to balance global coverage with fine-scale detail, especially as policymakers and planners increasingly need high-resolution data for regional adaptation strategies. Advances in high-performance computing (HPC) and machine learning are helping to address this challenge by enabling higher-resolution models and downscaling techniques that take the coarse outputs from GCMs and refine them to a local scale. Still, these techniques come with their own computational and technical demands.

Long-Term Projections and Nonlinear Feedbacks

Long-term climate projections predicting climate changes over centuries pose a significant challenge due to the nonlinear nature of the climate system and the potential for feedback mechanisms to alter future climate states in unpredictable ways [87]. Nonlinear feedbacks occur when small changes in one part of the system trigger larger, sometimes irreversible changes in another. These feedbacks are often difficult to predict and represent in models. One example is the ice-albedo feedback mentioned in earlier sections. As ice melts, the Earth's surface becomes less reflective (lower albedo), absorbing more heat and leading to further ice melt. This positive feedback loop accelerates warming, especially in polar regions. Another critical feedback is the release of methane from thawing permafrost. Methane is a much more potent greenhouse gas than CO2, and its release could significantly accelerate global warming, potentially pushing the climate system toward tipping points [88]. Tipping points are thresholds beyond which small perturbations can lead to large, potentially irreversible changes in the climate system. Examples include the potential collapse of the West Antarctic Ice Sheet, leading to several meters of sea-level rise, or the disruption of the Atlantic Meridional Overturning Circulation (AMOC), which would dramatically alter weather patterns in Europe and North America [89]. Accurately modeling these tipping points is incredibly challenging due to the complex interactions between different components of the Earth system and the inherent uncertainty in how these processes will unfold over time. Moreover, natural climate variability such as volcanic eruptions, solar cycles, and internal ocean–atmosphere dynamics like ENSO (El Niño-Southern Oscillation) adds another layer of complexity to long-term projections [90]. These factors can introduce noise into the climate signal, making it harder to isolate the effects of human activities and predict how the climate will evolve over centuries.

Computational Complexity and Model Interactions

The growing complexity of climate models particularly Earth System Models (ESMs) that integrate atmospheric, oceanic, land, and biogeochemical processes requires vast amounts of computational power [43]. These models simulate millions of interactions between different components of the Earth system, necessitating the use of supercomputers and parallel computing architectures to run simulations in a reasonable amount of time. However, even with access to cutting-edge technology, trade-offs must be made between model complexity, spatial resolution, and simulation time. As models become more detailed, the computational burden increases, often making it impossible to run simulations over the long timescales required for climate projections. For example, to model the climate up to the year 2100, researchers must often run multiple simulations with different initial conditions or emission scenarios to account for uncertainty. Each simulation may take weeks or even months of computer time, limiting the number of scenarios or the resolution at which models can be run. Furthermore, model coupling the integration of different sub-models for the atmosphere, ocean, land, and ice adds complexity. Each of these components operates on different spatial and temporal scales and coupling them together can lead to errors or inconsistencies. For example, atmospheric processes occur on much shorter timescales than oceanic processes, so climate models must account for these differences while ensuring that interactions between the components are accurately represented.

Model Validation and Data Limitations

Model validation is another critical challenge in climate modeling [91]. Figure 2 shows the steps involved in validating AI-driven climate models. Climate models are validated by comparing their output to observational data from sources like satellites, weather stations, ocean buoys, and paleoclimate records. While modern technology provides a wealth of data, observational gaps still exist, especially in remote regions such as the deep ocean, polar areas, and parts of the developing world. These gaps can lead to biases or uncertainties in model projections. In addition, many climate models rely on historical data to tune their parameters and improve accuracy. However, this data often spans only the past few decades, limiting the ability to capture long-term trends or validate models against rare climate events, such as major volcanic eruptions or multi-century droughts. Paleoclimate data, derived from ice cores, tree rings, and sediment records, can help extend the observational record, but these data sources come with their own uncertainties [92]. In summary, while climate models have made great strides in improving our understanding of the Earth's climate system, significant challenges remain. Uncertainty in climate sensitivity, resolution trade-offs, nonlinear feedbacks, computational complexity, and data limitations all pose obstacles to producing accurate, high-resolution, long-term projections. Overcoming these challenges requires continued advances in computational power, observational data collection, and modeling techniques, as well as interdisciplinary collaboration across climate science, computer science, and data analytics.

3 Predictive Climate Modeling Using AI

3.1 Introduction to AI in Climate Science

AI is transforming climate science by offering powerful tools for climate modeling, prediction, and data analysis. Traditional climate models, based on physical laws such as thermodynamics, fluid dynamics, and radiative transfer, have proven effective in simulating the Earth's climate system. However, these models face challenges in resolving fine-scale features, modeling nonlinear processes, and processing large datasets. AI addresses many of these limitations by leveraging ML algorithms, which excel at detecting patterns in vast datasets and identifying complex relationships that may be difficult to capture with physical equations alone. AI models bring several advantages to climate science. They are particularly useful for downscaling improving the spatial resolution of coarse global models to provide more detailed regional predictions and for parameterizing subgrid scale processes that are too small or complex to be directly simulated in traditional models. Moreover, AI techniques are proving invaluable in forecasting extreme weather events, which involve chaotic and nonlinear dynamics that traditional models struggle to predict with accuracy [93, 94]. AI-driven models can process high-dimensional data from sources like satellites, ground stations, and climate reanalysis products [95]. This allows them to find relationships between climate variables such as temperature, precipitation, wind patterns, and ocean currents that would be computationally infeasible to capture in purely physics-based models. As a result, AI models can complement traditional approaches by improving resolution, offering faster predictions, and integrating real-time data into climate forecasting systems [28].

Fig. 2 Flowchart of model validation processes

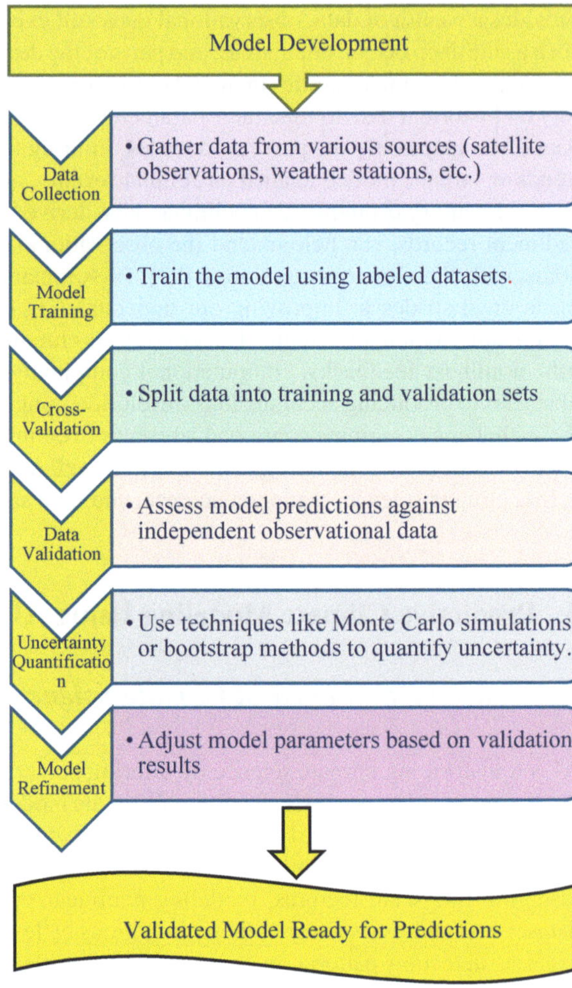

Complementing Traditional Climate Models

While traditional climate models simulate Earth's climate using physical laws, AI models rely on data-driven insights to make predictions. This distinction is critical because AI models do not necessarily need to understand the underlying physics to make accurate predictions; instead, they learn patterns and relationships directly from the data. This approach is especially useful in situations where the physics are too complex or poorly understood to model explicitly, such as cloud microphysics, aerosol interactions, or deep ocean dynamics. AI techniques can supplement traditional models by:

- Improving resolution: AI can downscale global climate model outputs to generate high-resolution regional predictions [96].

- Filling data gaps: AI techniques such as interpolation and extrapolation can fill in missing data points in climate datasets, especially in regions with sparse observations [97].
- Enhancing predictive accuracy: AI models can improve the accuracy of predictions by refining model parameters based on historical data, reducing bias, and optimizing forecasts for extreme events [50].

One area where AI has proven particularly effective is in post-processing climate model outputs [28]. For instance, after a traditional climate model has simulated future climate conditions, AI can refine these predictions by accounting for factors that are difficult to represent in physical models, such as local topography, land-use changes, or human activity. This process, known as statistical downscaling, allows AI to generate more accurate and actionable climate predictions at the local scale, which is crucial for urban planning, disaster preparedness, and agricultural management.

Handling Big Data in Climate Science

AI's ability to handle big data is one of its most transformative features for climate science [98]. As climate models and observational networks generate increasingly large and complex datasets, the challenge of processing and analyzing this data becomes more difficult. Traditional climate models often struggle to incorporate real-time observational data due to their reliance on computationally expensive simulations. In contrast, AI techniques such as machine learning and deep learning can quickly process multidimensional data and detect hidden patterns within it.

For example, AI algorithms can process data from sources like:

- Satellite observations: High-resolution data on sea surface temperatures, cloud cover, and atmospheric composition [99].
- Ground-based sensors: Weather stations providing continuous measurements of temperature, humidity, and precipitation [100, 101].
- Reanalysis products: Comprehensive climate datasets combining observations with model outputs to produce a consistent historical record of climate variables [102].

By analyzing these datasets, AI can uncover nonlinear relationships and spatial–temporal patterns that traditional models may miss. This capability is especially valuable for short-term weather forecasting and extreme event prediction, where real-time data integration is essential for accurate predictions.

AI for Extreme Weather Prediction

Another key advantage of AI is its ability to predict extreme weather events [103]. Traditional models often struggle to forecast events like hurricanes, tornadoes, heatwaves, and floods because these phenomena involve complex, nonlinear interactions between atmospheric, oceanic, and land processes. AI models, particularly those based on deep learning, can analyze vast amounts of historical data to identify patterns that precede extreme weather events, enabling more accurate and timely predictions, as shown in the steps of Fig. 3.

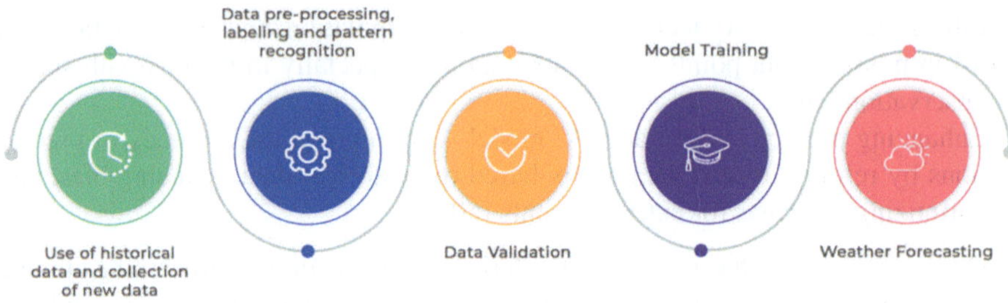

Fig. 3 Steps of whether forecasting using AI

For instance, AI techniques have been used to:

- Predict hurricane intensity and track landfall by analyzing real-time satellite data alongside historical hurricane data [104].
- Forecast heatwaves and droughts by detecting anomalies in temperature and precipitation patterns [105].
- Improve flood prediction models by analyzing river flow data, precipitation forecasts, and land cover information [106, 107].

AI is especially effective in forecasting the timing and location of extreme events, which can provide critical lead time for disaster preparedness and response efforts. These models are trained on vast datasets of past events, enabling them to recognize the early signs of an approaching extreme event and improve the accuracy of weather warnings. This has been further discussed in Sect. 4.4 of this chapter.

Bridging Data Gaps and Uncertainty

AI also plays a crucial role in addressing data gaps in climate science, particularly in developing regions or remote areas where observational networks are sparse [95]. Traditional climate models often suffer from data sparsity, making it difficult to generate reliable predictions for these areas. AI techniques, such as data interpolation, spatial extrapolation, and data augmentation, help bridge these gaps by generating plausible estimates based on available data. However, it's important to acknowledge that while AI can fill gaps in the data, it also introduces new challenges related to uncertainty. AI models are often seen as black boxes, meaning that their internal decision-making processes are not easily interpretable. This can lead to uncertainty in AI-based predictions, especially when the models are extrapolating data in regions with limited observations. For climate predictions to be trusted, it is essential that AI models incorporate physical constraints and provide transparent outputs that can be validated against real-world observations. To address this issue, researchers are increasingly turning to hybrid models that combine the strengths of AI with traditional physics-based approaches [108]. These models leverage AI's ability to handle large datasets and detect patterns, while ensuring that the predictions adhere to the known laws of physics, thus enhancing the interpretability and reliability of climate forecasts.

3.2 AI Techniques for Climate Prediction

AI has emerged as a transformative force in climate science, enabling researchers to enhance traditional climate models and improve predictive capabilities. This section delves into key AI techniques used in climate prediction, explaining their mechanisms, applications, and the reasons for their effectiveness.

Supervised Learning: Supervised learning techniques, such as linear regression, decision trees, and support vector machines (SVMs), are pivotal for forecasting climate variables like temperature and precipitation based on historical climate data [109]. These methods are effective because they can model complex relationships between input features (like atmospheric pressure, humidity, and sea surface temperatures) and target variables (such as future temperature or precipitation levels). The performance of supervised learning methods is highly dependent on the availability of large amounts of labeled data. For instance, when predicting temperature changes, the model requires extensive historical temperature records alongside relevant meteorological data to train effectively. The quality and quantity of this training data directly influence the accuracy of predictions; insufficient or noisy data can lead to poor model performance [110]. Real-world applications of supervised learning in climate science include predicting seasonal weather patterns and analyzing trends in long-term climate data. For example, SVMs have been employed to forecast extreme weather events by classifying atmospheric conditions that precede such occurrences.

Unsupervised Learning: Unsupervised learning techniques, including clustering algorithms like K-means, play a crucial role in identifying distinct climate patterns and anomalous weather events [111]. These algorithms group similar data points without predefined labels, allowing researchers to uncover underlying structures within complex datasets. In climate science, clustering can help identify regions with similar climatic characteristics or detect unusual weather phenomena that may indicate shifts in climate patterns. For example, K-means clustering has been used to categorize different phases of the El Niño-Southern Oscillation (ENSO), aiding in the understanding of its impacts on global weather systems [26, 112]. By recognizing these patterns, scientists can improve long-term forecasting models and enhance their ability to predict future climatic conditions based on historical data.

Deep Learning (DL): Deep learning models have revolutionized climate prediction by effectively addressing challenges associated with high-dimensional datasets and complex nonlinear relationships. Among these models, Convolutional Neural Networks (CNNs) and Recurrent Neural Networks (RNNs) have shown remarkable success in spatial and temporal predictions [113].

- Convolutional Neural Networks (CNNs): CNNs excel at processing grid-like data structures, making them ideal for downscaling global climate data to finer resolutions [114]. By capturing spatial hierarchies in the data, CNNs can generate more accurate regional climate projections. For instance, they have been applied to improve precipitation forecasts by translating coarse outputs from General Circulation Models (GCMs) into high-resolution predictions that are critical for local planning.

- Recurrent Neural Networks (RNNs): RNNs are particularly valuable for temporal predictions due to their ability to maintain information over time [115]. Long Short-Term Memory networks (LSTMs), a specialized type of RNN, are adept at capturing long-term dependencies in sequential data. They have been successfully utilized for tracking seasonal and decadal changes in climate variables. For example, LSTMs have been employed to predict long-term temperature trends by analyzing historical temperature records alongside relevant oceanic indices [116].

Neural Networks (NNs): Neural networks encompass various architectures that contribute significantly to climate prediction:

- Feedforward Neural Networks (FNNs): FNNs are effective for capturing nonlinear relationships within climate data [117]. They can model interactions between multiple variables such as sea surface temperatures and atmospheric pressure allowing for more nuanced predictions of climatic phenomena like hurricanes or monsoons.
- Long Short-Term Memory (LSTM): As previously mentioned, LSTMs are crucial for modeling long-term dependencies in climate data [118]. Their application extends to projecting ocean–atmosphere interactions over decades, providing insights into how these interactions influence global climate trends.
- Generative Adversarial Networks (GANs): GANs are powerful tools for generating synthetic datasets and filling gaps in missing data within climate projections [119, 120]. By learning from historical datasets, GANs can simulate alternative future scenarios based on various assumptions about greenhouse gas emissions or land-use changes. This capability is particularly valuable for exploring potential impacts of different climate policies or adaptation strategies.

AI techniques such as supervised learning, unsupervised learning, deep learning models like CNNs and LSTMs, along with neural networks including GANs, offer robust frameworks for enhancing climate prediction capabilities. Each method addresses specific challenges within the field ranging from capturing nonlinear relationships to generating synthetic datasets thus contributing significantly to our understanding of future climatic conditions. However, it is essential to acknowledge the limitations of these methods; supervised learning requires extensive labeled datasets while deep learning models often operate as black boxes that challenge interpretability and trustworthiness in predictions. Addressing these challenges will be crucial as we continue to integrate AI into climate science.

3.3 Hybrid AI Models

Hybrid AI models represent a significant advancement in climate prediction by integrating artificial intelligence techniques with traditional physics-based models [28]. This approach combines the strengths of both methodologies, allowing AI to leverage data-driven insights while ensuring that predictions adhere to established physical laws. Such integration is particularly vital in climate science, where the complexity of natural systems necessitates models that are both accurate and physically plausible.

Physics-Informed Neural Networks (PINNs): Physics-Informed Neural Networks (PINNs) are a notable example of hybrid models that embed physical laws directly into the learning process [121]. By incorporating fundamental principles such as conservation of energy and mass conservation, PINNs ensure that the predictions made by the neural network remain consistent with known physical constraints. This is critical when modeling complex processes like cloud microphysics or ocean dynamics, where traditional neural networks may struggle due to their "black-box" nature. For instance, in modeling atmospheric phenomena, PINNs can be designed to respect the Navier–Stokes equations governing fluid dynamics. By doing so, they not only improve the accuracy of predictions but also enhance interpretability, as users can trace how physical laws influence model outputs. This capability is essential for applications such as weather forecasting and climate simulations, where understanding the underlying mechanisms is crucial for decision-making.

Ensemble Learning: Ensemble learning techniques, including methods like random forests and boosting, are another powerful component of hybrid AI models [122]. These techniques aggregate predictions from multiple models to improve overall predictive accuracy and robustness while reducing the risk of overfitting. In climate modeling, ensemble methods are particularly useful for quantifying uncertainty by exploring a range of possible future scenarios. For example, ensemble learning can be employed to combine outputs from various climate models both AI-driven and traditional physics-based models to produce a more comprehensive forecast. This approach allows researchers to capture a wider array of potential outcomes, providing policymakers with valuable insights into the range of possible climate futures. By quantifying uncertainty through ensemble predictions, scientists can better inform strategies for climate adaptation and mitigation.

Real-World Applications: Hybrid AI models have shown promise in several real-world applications within climate science:

- Downscaling: PINNs have been utilized for downscaling coarse-resolution climate model outputs to finer spatial resolutions. By embedding physical constraints into the learning process, these models generate high-resolution climate projections that are critical for local impact assessments.
- Weather Prediction: Ensemble learning techniques have been effectively applied in weather prediction contexts. For instance, combining outputs from different forecasting models can enhance the reliability of predictions regarding extreme weather events like hurricanes or heatwaves.
- Long-Term Climate Forecasts: Hybrid approaches that integrate AI with traditional ocean circulation models have improved long-term sea-level rise projections. By accounting for complex interactions between atmospheric and oceanic processes, these models provide more accurate forecasts essential for coastal planning and management.

The hybrid AI models represent a promising frontier in climate prediction by merging data-driven AI techniques with established physical principles. By enhancing model interpretability and robustness while addressing complex climate processes, these hybrid approaches pave the way for more accurate and reliable climate forecasts that can inform critical decisions in an era of rapid environmental change.

3.4 Data Requirements and Challenges

Data is the lifeblood of AI models, particularly in the realm of climate science, where vast and varied datasets are essential for accurate modeling and prediction. However, several challenges complicate the effective use of data in AI-driven climate models:

Big Data in Climate Science: AI-based climate models rely on extensive datasets sourced from satellite observations, ocean buoys, weather stations, and reanalysis products [123]. These datasets provide the high-resolution inputs necessary for training AI models effectively. For example, satellite data can offer insights into atmospheric conditions over large areas, while ground-based observations provide localized information that is critical for understanding microclimates. The integration of these diverse data sources enhances model accuracy but also increases the complexity of data management.

Heterogeneous Data Sources: Climate data is inherently heterogeneous, coming in multiple formats and resolutions [124]. This diversity presents significant challenges for integration. Remote sensing data, such as that obtained from satellites, typically offers high spatial resolution but may suffer from low temporal resolution due to infrequent overpasses. Conversely, station data provides high temporal resolution but often has limited spatial coverage. This mismatch complicates the development of comprehensive models that require consistent and uniform datasets across various geographical regions and time scales.

Data Availability in Developing Regions: While developed regions benefit from abundant climate data, many developing regions face challenges related to sparse and incomplete climate observations. This lack of data can hinder accurate modeling and forecasting efforts in these areas. To address this issue, AI-based data interpolation techniques can be employed to fill gaps in observational records. However, the accuracy of these interpolations remains a concern, as they depend heavily on the underlying assumptions made during the modeling process and the quality of available data.

Computational Complexity: Training deep learning models on high-resolution climate datasets is computationally demanding [125]. These models often require high-performance computing resources or cloud-based solutions to scale effectively. The need for significant computational power can limit accessibility for researchers in resource-constrained environments and may pose barriers to widespread adoption of advanced AI techniques in climate science.

Model Interpretability: One of the greatest challenges associated with AI in climate science is the black-box nature of many AI models, particularly deep learning approaches [126]. While these models can achieve remarkable predictive accuracy, their lack of transparency raises concerns about trustworthiness and interpretability. Ensuring that AI-based predictions adhere to established physical principles is crucial for maintaining confidence in climate forecasts. Researchers must develop methods to enhance model interpretability, allowing stakeholders to understand how predictions are derived and how they relate to underlying physical processes. In conclusion, while the integration of AI into climate modeling presents significant opportunities for enhancing predictive capabilities, it also introduces a range of challenges related to data requirements and model complexity. Addressing these challenges will be essential for maximizing the potential of AI-driven approaches in understanding and mitigating climate change impacts.

4 AI-Driven Climate Prediction Models: Case Studies

4.1 Climate Downscaling Using Neural Networks

In climate science, downscaling refers to the process of improving the spatial resolution of outputs from General Circulation Models (GCMs) to produce finer-scale climate projections that are more relevant for local and regional applications. AI-driven downscaling approaches, particularly those utilizing CNNs, have emerged as powerful tools for this purpose. By training on high-resolution observational data, CNNs can generate detailed regional predictions from the coarse outputs of GCMs, significantly enhancing the granularity and applicability of climate forecasts.

The Mechanism of Downscaling with CNNs

CNNs are particularly well-suited for downscaling tasks due to their ability to capture spatial hierarchies in data [127]. They process, input data through multiple layers of convolutional filters that detect patterns and features at various scales. In the context of climate downscaling, CNNs can learn complex relationships between large-scale atmospheric conditions and localized climate phenomena, such as precipitation patterns or temperature variations. For instance, a study focused on downscaling precipitation data over Europe demonstrated that CNNs could effectively translate coarse GCM outputs into high-resolution predictions [128]. By training on extensive historical datasets that include both satellite observations and ground-based measurements, these models can learn to identify regional climatic features that GCMs may overlook due to their lower resolution. This capability allows for more accurate assessments of local climate impacts, which are critical for water resource management, agriculture, and urban planning. The flow chart in Fig. 4 shows the downscaling process using CNNs.

Applications Across Regions

The application of AI-driven downscaling techniques has been observed in various regions, including North America and Europe. In South America, CNNs have been employed to improve temperature and precipitation forecasts for specific states or provinces, providing valuable insights for local governments and industries reliant on accurate climate data [129]. Similarly, in Europe, researchers have utilized these models to assess the potential impacts of climate change on regional ecosystems and biodiversity [130].

Advantages Over Traditional Methods

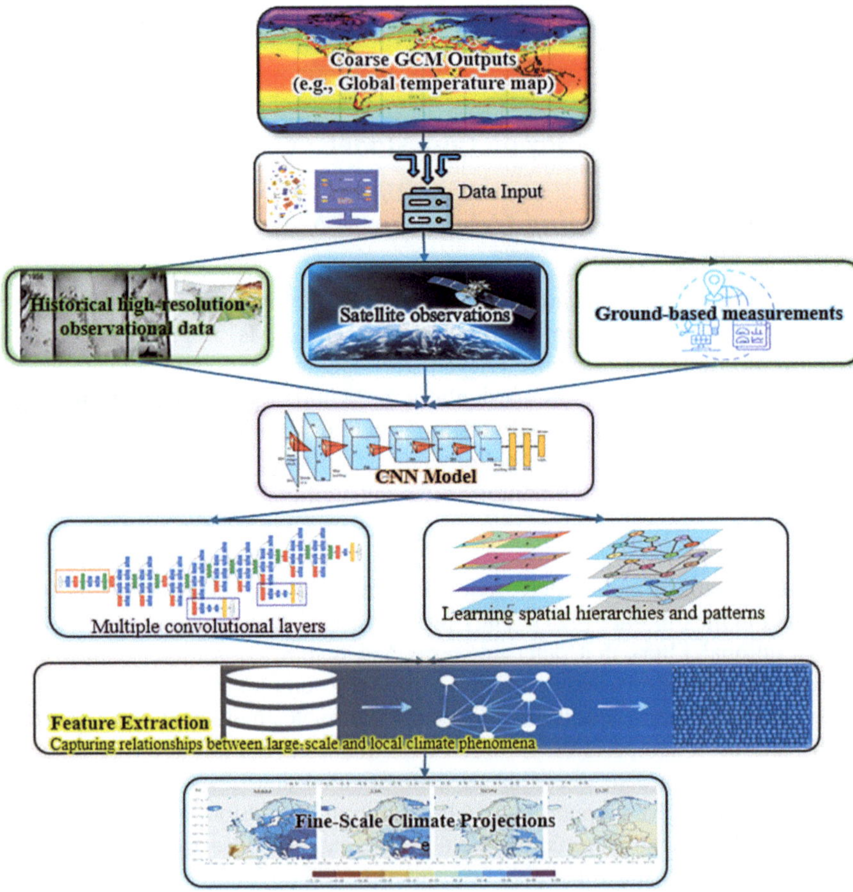

Fig. 4 This flowchart shows the downscaling process using CNNs

AI-driven downscaling approaches offer several advantages over traditional statistical or dynamical downscaling methods:

- Higher Accuracy: CNNs can leverage vast amounts of data to improve predictive accuracy by learning from complex patterns that traditional methods may not capture.
- Efficiency: Once trained, CNNs can generate predictions much faster than traditional models, making them suitable for real-time applications.
- Flexibility: These models can be adapted to various climatic variables and geographical regions, allowing for widespread applicability across different contexts.

Challenges and Future Directions

Despite their promise, the implementation of AI-driven downscaling techniques is not without challenges. The requirement for high-quality training data is paramount; insufficient or biased datasets can lead to inaccurate predictions. Moreover, the black-box nature of neural networks raises concerns regarding interpretability, understanding how specific inputs influence model outputs is crucial for stakeholder trust. Future, research should focus on enhancing model interpretability and incorporating physical constraints into AI-driven downscaling approaches. By developing hybrid models that combine AI techniques with established physical principles, researchers can create more robust and reliable climate projections that adhere to known scientific laws. In summary, AI-driven downscaling using neural networks represents a significant advancement in climate modeling. By providing high-resolution regional projections from coarse GCM outputs, these techniques enhance our understanding of localized climate impacts and inform decision-making processes across various sectors. As technology continues to evolve, the integration of AI in climate science will likely yield even more sophisticated tools for addressing the challenges posed by climate change.

4.2 Predicting El Niño Events with Machine Learning

The El Niño-Southern Oscillation (ENSO) is one of the most significant drivers of global climate variability, influencing weather patterns, agricultural productivity, and ecosystem dynamics worldwide [131, 132]. Accurate prediction of El Niño events is crucial for mitigating their impacts and enhancing preparedness across various sectors. In recent years, machine learning (ML) models have been increasingly employed to forecast the occurrence of El Niño events months in advance, surpassing the capabilities of traditional physical models in certain contexts.

Mechanisms of Prediction

Machine learning models, including Support Vector Machines (SVMs) and deep neural networks, are particularly effective in predicting El Niño events due to their ability to analyze complex relationships within large datasets [133]. These models incorporate a range of oceanic and atmospheric indices, such as sea surface temperatures (SSTs), wind patterns, and pressure anomalies, which are critical indicators of ENSO phases. For instance, SVMs can classify different phases of ENSO by learning from historical data that includes both successful predictions and instances where predictions failed. By identifying patterns that precede El Niño events, these models can provide early warnings that are essential for agricultural planning and disaster risk management [134, 135].

Advantages Over Traditional Models

The application of ML techniques for predicting El Niño offers several advantages over traditional physical models:

Improved Accuracy: ML models can capture nonlinear relationships and interactions that may be difficult for traditional models to represent. This capability allows them to identify subtle signals in the data that indicate an impending El Niño event.

Timeliness: Machine learning algorithms can process vast amounts of data quickly, enabling timely predictions that can be crucial for stakeholders needing to prepare for potential impacts.

Adaptability: These models can continuously learn from new data, allowing them to adapt to changing climate conditions and improve their predictive performance over time.

Case Studies and Real-World Applications

Several studies have demonstrated the effectiveness of machine learning in predicting El Niño events. For example, researchers have successfully employed deep learning techniques to analyze satellite-derived SST data alongside atmospheric pressure measurements [136, 137]. These models have shown promise in providing accurate forecasts several months in advance, which is beneficial for sectors such as agriculture and fisheries that are directly affected by ENSO dynamics [138]. In another case, ensemble approaches combining multiple machine learning algorithms have been utilized to enhance prediction reliability. By aggregating outputs from different models, researchers can produce a consensus forecast that accounts for uncertainties inherent in individual predictions.

Challenges and Future Directions

Despite the advancements in using machine learning for predicting El Niño events, challenges remain. The reliance on historical data means that model performance can be affected by shifts in climate patterns or anomalies not previously encountered. Additionally, the interpretability of complex ML models poses challenges; stakeholders require transparent insights into how predictions are made to build trust in these new methodologies. Future research should focus on improving model interpretability while enhancing predictive accuracy through hybrid approaches that integrate physical principles with machine learning techniques. By addressing these challenges, researchers can develop more robust forecasting systems that effectively predict El Niño events and their associated impacts on global climate variability. In conclusion, machine learning represents a promising frontier in predicting El Niño events. By leveraging large datasets and advanced algorithms, these approaches enhance our ability to anticipate significant climate phenomena, thereby facilitating better preparedness and response strategies across various sectors impacted by ENSO dynamics.

4.3 Hybrid AI for Sea-Level Rise Prediction

Predicting future sea-level rise is a critical challenge in climate science, with significant implications for coastal communities and ecosystems worldwide [139]. Traditional models based on ocean circulation physics have provided valuable insights, but they often struggle to capture the complex, nonlinear interactions between atmospheric and oceanic processes that drive sea-level changes. To address this challenge, researchers have turned to hybrid approaches that integrate physics-based ocean models with AI-driven analysis of observational data, particularly satellite altimetry measurements.

Leveraging Satellite Data with AI

Satellite altimetry missions, such as TOPEX/Poseidon, Jason-1, and Jason-2, have provided a wealth of data on sea surface height (SSH) over the past few decades [140]. This data, when combined with in-situ measurements from tide gauges and Argo floats, offers a comprehensive view of global and regional sea-level trends. However, extracting meaningful insights from these vast datasets requires advanced analytical techniques. Machine learning algorithms, particularly neural networks, have proven effective in processing and interpreting satellite altimetry data. By training on historical SSH measurements alongside relevant climate indices (e.g., ocean temperature, wind patterns, atmospheric pressure), AI models can identify complex patterns and relationships that drive sea-level variability [141]. This allows for more accurate projections of future sea-level rise under various climate change scenarios.

Integrating Physics-Based Models

While AI techniques excel at pattern recognition and prediction, they lack the fundamental understanding of physical processes that traditional ocean circulation models provide. Hybrid approaches bridge this gap by combining the strengths of both methodologies. In a hybrid framework, AI models are used to enhance or downscale outputs from physics-based ocean models, such as the Parallel Ocean Program (POP) or the Modular Ocean Model (MOM) [142]. By learning from the discrepancies between model outputs and observational data, AI algorithms can identify systematic biases or missing processes in the physical models. This information is then used to improve model parameterizations or to generate high-resolution regional projections.

Advantages and Applications

Hybrid AI models for sea-level rise prediction offer several advantages over traditional approaches:

Improved Accuracy: By accounting for nonlinear interactions and leveraging vast datasets, hybrid models can provide more reliable projections of future sea-level rise.

Enhanced Resolution: AI techniques enable the downscaling of coarse-resolution ocean models to finer spatial scales, allowing for more detailed assessments of local impacts.

Computational Efficiency: Once trained, AI models can generate predictions rapidly, enabling real-time monitoring and forecasting of sea-level changes.

These hybrid approaches have been applied in various contexts, such as assessing the potential impacts of sea-level rise on coastal cities, evaluating the effectiveness of adaptation strategies like seawalls or beach nourishment, and informing decisions related to infrastructure planning and ecosystem management.

Challenges and Future Directions

Despite the promise of hybrid AI models, several challenges remain. Ensuring the interpretability and trustworthiness of these models is crucial for stakeholder acceptance and policy decisions. Additionally, the reliance on historical data means that hybrid models may struggle to account for novel or unprecedented climate scenarios, such as the potential collapse of ice sheets or the release of methane from thawing permafrost. Future research should focus on developing more robust and transparent hybrid frameworks that seamlessly integrate physical principles with AI techniques. This may involve the use of physics-informed neural networks or other hybrid architectures that explicitly incorporate conservation laws and boundary conditions. Additionally, ensemble approaches that combine multiple hybrid models can help quantify uncertainty and provide a more comprehensive assessment of future sea-level rise risks. A review study by Bahari et al., showed that hybrid AI models represent a promising avenue for predicting future sea-level rise [143]. By leveraging the strengths of both data-driven and physics-based approaches, these models offer improved accuracy and resolution compared to traditional methods. As climate change continues to drive sea-level rise, hybrid AI models will play an increasingly important role in informing adaptation strategies and protecting vulnerable coastal communities worldwide.

4.4 Extreme Weather Event Prediction

AI techniques are increasingly being employed to predict extreme weather events, including hurricanes, droughts, and heatwaves [144]. These events pose significant risks to human life, infrastructure, and ecosystems, making accurate forecasting essential for preparedness and response strategies. Recent advancements in deep learning models have demonstrated their effectiveness in processing large-scale datasets from satellites, radars, and climate reanalyzes to improve predictions of storm intensity and landfall locations.

Mechanisms of Prediction

Deep learning models, particularly CNNs and RNNs, have been utilized to analyze vast amounts of data related to atmospheric conditions. By training on historical records of extreme weather events alongside real-time observational data, these models can learn complex patterns that precede such occurrences. For instance, CNNs can effectively process spatial data from satellite imagery to identify features indicative of developing storms, while RNNs can capture temporal dependencies in weather patterns over time [145–147].

Improved Accuracy Compared to Traditional Methods

The integration of AI into extreme weather prediction has led to significant improvements in accuracy over traditional forecasting methods. Traditional models often rely on physical equations and historical averages, which may not adequately capture the dynamic nature of extreme weather events. In contrast, AI-driven approaches can adaptively learn from new data, allowing for more precise predictions that account for the unique characteristics of each event. For example, studies have shown that deep learning models can predict hurricane intensity and track their paths with greater accuracy than conventional numerical weather prediction models [148, 149]. By analyzing a combination of satellite data, radar observations, and climate reanalyzes, these AI models provide timely forecasts that are critical for emergency management and public safety.

Challenges in Predicting Extreme Events

Despite the advancements in AI-driven prediction techniques, challenges remain in forecasting extreme weather events:

- Low Frequency and High-Impact Nature: Extreme weather events are relatively rare compared to normal weather patterns. This low frequency makes it difficult for models to learn effectively from historical data, leading to potential biases in predictions.
- Imbalanced Datasets: The underrepresentation of extreme events in training datasets can result in models that are less sensitive to these occurrences. This imbalance may lead to overconfidence in predictions for more common weather patterns while underestimating the likelihood or severity of extreme events.

Addressing Challenges with Techniques

To mitigate the challenges associated with imbalanced datasets, several techniques have been employed:

- Data Augmentation: This approach involves artificially increasing the size of training datasets by generating synthetic examples of extreme weather events. For instance, researchers can use techniques like oversampling or generating variations of existing data points to create a more balanced dataset.
- Ensemble Learning: Ensemble methods combine predictions from multiple models to improve overall accuracy and robustness [150]. By aggregating outputs from various AI algorithms such as CNNs and RNNs ensemble learning can enhance the reliability of forecasts for extreme weather events by providing a consensus prediction that accounts for uncertainties.

Future Directions

As research continues to advance in the field of AI-driven extreme weather prediction, future efforts should focus on improving model interpretability and transparency. Understanding how AI models arrive at their predictions is crucial for building trust among stakeholders and ensuring that forecasts are actionable. Additionally, integrating AI techniques with traditional physical models could enhance predictive capabilities by combining data-driven insights with established scientific principles. This hybrid approach may help address some limitations inherent in purely data-driven methods while providing a more comprehensive understanding of extreme weather dynamics. In conclusion, AI techniques are transforming the landscape of extreme weather event prediction by leveraging large-scale datasets and advanced modeling approaches. While challenges remain particularly concerning data imbalances ongoing advancements in machine learning hold great promise for improving our ability to forecast severe weather events accurately. As these technologies continue to evolve, they will play an increasingly important role in enhancing resilience against climate-related hazards.

5 Ethical and Policy Implications of AI in Climate Prediction

The integration of AI into climate science presents significant ethical and policy challenges that must be carefully considered as these technologies become more prevalent in informing critical decision-making processes [151]. AI-driven models are increasingly utilized to guide policy decisions related to infrastructure planning, disaster preparedness, and greenhouse gas (GHG) mitigation strategies [152, 153]. However, the complexities associated with these models raise important questions regarding their reliability, transparency, and potential biases.

Interpretability and Transparency

One of the foremost concerns regarding AI models in climate prediction is their interpretability. Policymakers require outputs that are not only accurate but also transparent and explainable. The "black-box" nature of many AI algorithms, particularly deep learning models, complicates this requirement. When decisions are made based on model predictions such as allocating resources for disaster response or investing in climate resilience measures, stakeholders must understand how these predictions were generated. Without clear explanations of the underlying processes and assumptions, there is a risk that policymakers may misinterpret the information or make decisions based on flawed premises.

Potential Biases in AI Models

Another critical issue is the potential for biases within AI models, which can arise from various sources, including data gaps, historical inequalities, and model assumptions [154]. For example, if training datasets are not representative of all geographical regions or demographic groups, the resulting predictions may favor certain areas while neglecting others. This bias can lead to inequitable outcomes in policy decisions, particularly for vulnerable populations that may already be disproportionately affected by climate change. Moreover, biases stemming from model assumptions can result in flawed predictions with far-reaching consequences. For instance, if an AI model underrepresents the likelihood of extreme weather events due to historical data limitations, it may lead to insufficient preparedness measures being implemented in regions at risk [155].

Ethical Considerations in Policy Implementation

The ethical implications of using AI-driven predictions extend beyond model performance; they also encompass the broader societal impacts of policy decisions informed by these models. Policymakers must consider how their decisions affect different communities and ensure that vulnerable populations are not further marginalized. This requires commitment to equity and inclusivity in both the development of AI models and the implementation of policies based on their outputs (Fig. 5).

Recommendations for Ethical AI Use in Climate Prediction

To address these ethical and policy implications effectively, several recommendations can be made:

Enhancing Model Interpretability: Researchers should prioritize developing methods that enhance the transparency of AI models. Techniques such as explainable AI (XAI) can help elucidate how models arrive at their predictions, making it easier for policymakers to understand and trust the outputs [156].

Addressing Data Biases: Efforts should be made to ensure that training datasets are comprehensive and representative of diverse geographical and demographic contexts. This may involve actively seeking out underrepresented data or employing techniques like data augmentation to create more balanced datasets.

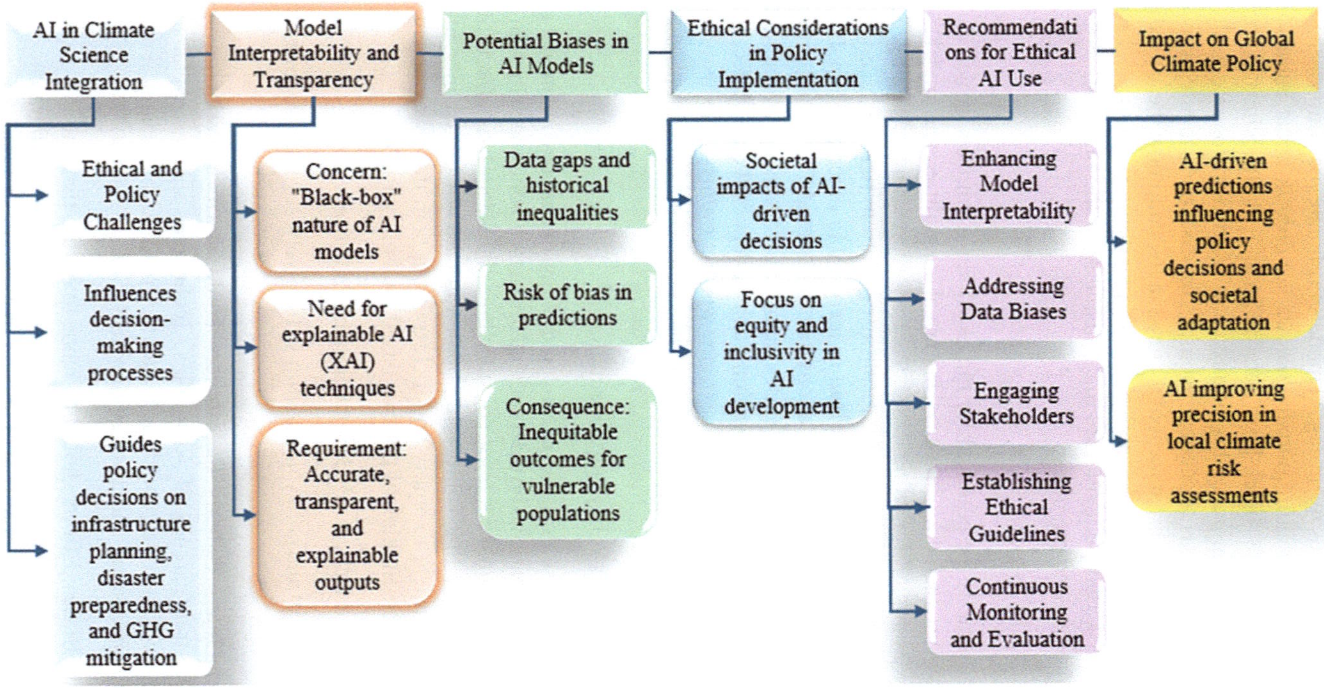

Fig. 5 This flowchart visually represents how AI's role in climate prediction is interconnected with ethical and policy considerations, leading to strategies that can ensure responsible and equitable use of AI technology

Engaging Stakeholders: Policymakers should engage with a wide range of stakeholders including scientists, community leaders, and affected populations—when developing climate policies informed by AI predictions. This collaborative approach can help identify potential biases and ensure that policies are equitable.

Establishing Ethical Guidelines: The establishment of ethical guidelines for the use of AI in climate science is essential. These guidelines should address issues related to data privacy, bias mitigation, transparency, and accountability to ensure responsible use of AI technologies.

Continuous Monitoring and Evaluation: Policymakers should implement mechanisms for continuous monitoring and evaluation of AI-driven predictions and their impacts on communities [157]. This feedback loop can help identify unintended consequences early on and allow for adjustments to be made as necessary.

Analysis on the Impact of AI-Driven Predictions on Policy Decisions and Societal Adaptation Measures

AI-driven climate predictions have the potential to dramatically influence policy decisions, shaping strategies for climate adaptation and mitigation at various levels of governance. The precision and granularity of AI-enhanced models provide policymakers with more detailed insights into regional and local climate impacts, enabling targeted responses. However, this growing reliance on AI-generated forecasts also raises critical ethical and practical concerns regarding policy formulation, equity, and societal adaptation.

Enhanced Precision in Policy Design

AI's ability to generate more localized and precise predictions allows governments to create highly tailored policy interventions. For example, by using AI to downscale climate models, regional authorities can forecast the effects of rising sea levels, urban heat islands, or changing precipitation patterns with greater accuracy. This granularity aids in infrastructure planning, disaster preparedness, and resource allocation, ensuring that responses are well-suited to the specific climate risks faced by different communities. For example, a coastal city receiving AI-based predictions of accelerated sea-level rise can proactively invest in flood defenses or relocate critical infrastructure, rather than reacting to more generalized predictions that might have delayed such investments [158].

Informing Long-Term Adaptation Strategies

AI-driven climate models also play a crucial role in shaping long-term adaptation strategies. Governments can use these predictions to create frameworks for sustainable agriculture, water resource management, and urban development that account for future climate variability [159]. These long-term strategies are essential for building societal resilience to climate change, particularly in vulnerable regions that lack the resources to adapt without external support. For example, AI models predicting prolonged droughts in arid regions could inform water conservation policies, crop selection for agriculture, or investment in desalination plants [160, 161].

This forward-looking approach ensures that societies are better equipped to manage the social and economic stresses caused by climate impacts.

Ethical Concerns Regarding Equity

While AI can significantly improve the accuracy of climate predictions, there is an ethical concern about how these predictions are utilized in policymaking, particularly regarding equity. The most vulnerable populations, those in developing countries, marginalized communities, or areas with less access to technological infrastructure may not benefit equally from the advancements in AI-driven predictions. In some cases, AI-based climate adaptation measures may prioritize wealthier regions or urban centers where data availability and infrastructure are better developed, exacerbating existing inequalities. Policymakers need to ensure that adaptation strategies are inclusive and prioritize those who are most at risk, rather than reinforcing disparities. For example, a city might use AI predictions to improve climate resilience in affluent neighborhoods by enhancing flood control measures, while poorer, less politically connected communities might continue to face inadequate protection from the same climate risks [162].

Trust and Transparency in Policy

Another critical aspect of AI-driven climate predictions is their inherent "black box" nature, which can undermine trust in policy decisions based on these models. For AI to meaningfully influence societal adaptation, policymakers must ensure that the models' outputs are transparent and interpretable. Without this, there is a risk of public distrust, which can hamper the implementation of essential adaptation measures. For example, a government might base significant climate adaptation investments on AI-driven predictions of extreme heatwaves or droughts [163, 164]. If the public cannot understand how these predictions were made or why certain decisions follow, it could lead to resistance against necessary changes in water management, urban planning, or energy usage. To mitigate this risk, hybrid AI models that combine traditional physics-based approaches with machine learning techniques offer a more interpretable framework, which can improve both trust and transparency. Clear communication from scientists and policymakers regarding how AI predictions are derived, the degree of uncertainty involved, and their implications for policy decisions is essential to maintaining public confidence.

Impact on Global Climate Policy

At the international level, AI-enhanced climate models provide valuable tools for setting more accurate and ambitious global climate targets. By providing more detailed projections of emissions scenarios and their likely outcomes, AI can inform global agreements like the Paris Accord and influence national commitments to reduce greenhouse gas emissions [165]. Countries can develop emissions reduction pathways that are based on more accurate assessments of regional vulnerabilities and risks, leading to more effective and equitable climate action globally. For example, a nation could leverage AI-based regional climate projections to set more realistic targets for reducing agricultural emissions while also developing adaptive measures for farmers in drought-prone areas [166].

AI-Driven Risk Management for Societal Adaptation

In the context of societal adaptation, AI can enhance the ability of governments and institutions to manage risks related to extreme weather events. For example, AI-based early warning systems can offer better predictions of floods, storms, and heatwaves, giving communities more time to prepare and reduce the impact on lives and livelihoods. This is particularly important in areas where traditional weather prediction models are less effective due to the complexity of local geographies or data limitations.

For example. AI-enhanced flood prediction systems could be deployed in regions with historically poor flood preparedness, providing local authorities with accurate early warnings that could help save lives and protect property [167, 168].

Adaptation for Climate-Sensitive Sectors

AI models also offer opportunities to optimize adaptation in climate-sensitive sectors such as agriculture, energy, and health. For instance, AI-based predictions of changing rainfall patterns or temperature fluctuations can inform agricultural practices, reducing the risk of crop failure. Similarly, AI can enhance the resilience of energy systems by optimizing the integration of renewable energy sources based on predictions of wind and solar availability. For example, in agriculture, AI-driven models that predict shifts in seasonal weather patterns can help farmers adopt climate-smart agricultural practices such as crop rotation, irrigation management, or soil conservation, increasing food security in the face of climate change [169].

By integrating AI-driven climate predictions into policy and societal adaptation strategies, governments can enhance precision, reduce risks, and improve the effectiveness of climate resilience efforts. However, the deployment of AI in this field requires careful consideration of ethical concerns, equity, transparency, and public trust. Ensuring that the benefits of AI-based climate predictions are shared broadly and that they inform inclusive policies is essential for addressing the global challenge of climate change. In conclusion, while the integration of AI into climate prediction holds great promise for enhancing our understanding of climate dynamics and informing effective policy responses, it also raises significant ethical and policy challenges. By prioritizing interpretability, addressing biases, engaging stakeholders, establishing ethical guidelines, and implementing continuous evaluation processes, we can harness the power of AI responsibly to create equitable climate policies that benefit all members of society.

6 Model Validation, Uncertainty Quantification, and Future Directions

The integration of artificial intelligence into climate modeling necessitates rigorous validation and uncertainty quantification to ensure the accuracy and reliability of predictions. As AI-driven models become more prevalent in informing climate-related decisions, understanding their performance and limitations is critical for stakeholders. This section discusses model validation techniques, approaches to uncertainty quantification, and future research directions in AI-driven climate science.

6.1 Model Validation Techniques

Validating AI-driven climate models is essential for establishing their credibility and ensuring that they produce reliable forecasts. Several techniques can be employed for model validation:

Cross-Validation: This statistical method involves partitioning the dataset into subsets, training the model on a portion of the data, and validating it on the remaining data. Cross-validation helps to assess how well the model generalizes to unseen data and reduces the risk of overfitting [170].

Independent Observational Datasets: Comparing model outputs against independent observational datasets is crucial for validating AI models. This comparison allows researchers to evaluate the accuracy of predictions and identify any systematic biases in the model's outputs.

Monte Carlo Simulations: This technique involves running simulations multiple times with varying input parameters to assess how uncertainty in inputs affects model outputs. By generating a distribution of possible outcomes, Monte Carlo simulations provide insights into the robustness of predictions [171].

Bootstrap Methods: Bootstrapping is a resampling technique that allows for estimating the distribution of statistics by repeatedly sampling from the dataset with replacement [172]. This method can be useful for quantifying uncertainty in model predictions by assessing variability in the results.

6.2 Uncertainty Quantification

AI models, particularly deep learning approaches, often lack built-in mechanisms for uncertainty quantification, which is especially problematic in climate science where uncertainty bounds are crucial for decision-making. Several approaches can be employed to estimate uncertainty in AI-driven climate predictions:

Ensemble Modeling: By combining predictions from multiple models or different runs of the same model, ensemble methods can provide a range of possible outcomes that reflect uncertainty. This approach helps capture variability in predictions and enhances reliability.

Bayesian Neural Networks: These networks incorporate Bayesian principles to estimate uncertainty by treating weights as distributions rather than fixed values [173]. This allows for probabilistic predictions that account for uncertainty in model parameters.

Gaussian Processes: Gaussian processes are a non-parametric approach to modeling that can provide uncertainty estimates along with predictions [174]. They are particularly useful for interpolation tasks and can quantify confidence intervals around predicted values.

6.3 Future Research Directions

The future of AI in climate science lies in developing hybrid models that integrate the strengths of both AI techniques and traditional physics-based approaches. As computational power continues to grow, AI models will likely improve in accuracy and scalability. However, several challenges must be addressed to fully realize the potential of AI in climate modeling:

Data Sparsity: Many regions lack sufficient observational data, which can hinder model training and validation. Future research should focus on developing methods to interpolate or augment datasets effectively.

Computational Demand: Training complex AI models on high-resolution climate datasets requires significant computational resources. Advancements in high-performance computing (HPC) and cloud-based solutions will be essential for scaling these models.

Model Interpretability: Enhancing the interpretability of AI models remains a critical challenge. Researchers should explore techniques that make model outputs more transparent and understandable to stakeholders, ensuring trust in AI-driven predictions. In conclusion, rigorous model validation and effective uncertainty quantification are vital components of integrating AI into climate prediction. By employing robust validation techniques and exploring innovative methods for estimating uncertainty, researchers can enhance the reliability of AI-driven climate models. As technology advances, addressing challenges related to data availability, computational demands, and interpretability will be crucial for maximizing the potential of AI in understanding and mitigating climate change impacts.

7 Conclusion

This chapter has explored the evolution of climate modeling, highlighting the transformative role of artificial intelligence in advancing climate science. While traditional climate models, grounded in physics-based approaches, have laid the foundation for our understanding of climate dynamics, they face inherent limitations in capturing the complex, nonlinear interactions within the Earth system. AI, with its capacity to process vast amounts of data and detect intricate patterns, offers a complementary approach, enhancing the resolution, speed, and accuracy of climate predictions. AI techniques like supervised learning, unsupervised learning, deep learning architectures such as CNNs and RNNs, and the emergence of hybrid models that integrate AI with physics-based approaches are revolutionizing climate science. These advancements enable more precise downscaling of global climate model outputs to regional scales, provide valuable insights for predicting extreme weather events, and offer a more comprehensive understanding of the carbon cycle and its feedbacks on the climate system. However, the integration of AI into climate prediction also presents a range of challenges that require careful consideration. Data requirements, computational complexity, and ensuring the interpretability of AI models are crucial areas of focus. Moreover, addressing ethical implications related to potential biases in AI models and ensuring equitable access to the benefits of AI-driven climate predictions are paramount for building trust and facilitating effective policy decisions. Future research directions should prioritize the development of hybrid models that seamlessly integrate physical principles with AI techniques, enhance model interpretability, and address data gaps, particularly in regions with limited observational coverage. Furthermore, establishing ethical guidelines for the use of AI in climate science and implementing continuous monitoring and evaluation processes are essential for ensuring the responsible and equitable application of these powerful technologies. By embracing these advancements while addressing the associated challenges, AI-driven climate prediction has the potential to significantly improve our ability to understand, predict, and respond to the complex challenges posed by climate change. Through collaborative efforts among scientists, policymakers, and stakeholders, AI can serve as a powerful tool for building a more resilient and sustainable future for all.

References

1. Collins, M., et al. (2013). *Long-term climate change: projections, commitments and irreversibility.*
2. Bates, J. R. (2016). Estimating climate sensitivity using two-zone energy balance models. *Earth and Space Science, 3*(5), 207–225.
3. Farzaneh, M. R., et al. (2012). An appropriate general circulation model (GCM) to investigate climate change impact. *International Journal of Hydrology Science and Technology, 2*(1), 34–47.

4. Flato, G. M. (2011). Earth system models: An overview. *Wiley Interdisciplinary Reviews: Climate Change, 2*(6), 783–800.
5. Change, I. P. O. C. (2007). Climate change 2007: The physical science basis. *Agenda, 6*(07), 333.
6. Trenberth, K. E., Fasullo, J. T., & Shepherd, T. G. (2015). Attribution of climate extreme events. *Nature Climate Change, 5*(8), 725–730.
7. Flato, G., et al. (2014). Evaluation of climate models. In *Climate change 2013: the physical science basis. Contribution of Working Group I to the Fifth Assessment Report of the Intergovernmental Panel on Climate Change* (pp. 741–866). Cambridge University Press.
8. Goodwin, P., Williams, R. G., & Ridgwell, A. (2015). Sensitivity of climate to cumulative carbon emissions due to compensation of ocean heat and carbon uptake. *Nature Geoscience, 8*(1), 29–34.
9. Meehl, G. A., et al. (2005). *How much more global warming and sea level rise? Science, 307*(5716), 1769–1772.
10. Riahi, K., et al. (2017). The Shared Socioeconomic Pathways and their energy, land use, and greenhouse gas emissions implications: An overview. *Global Environmental Change, 42*, 153–168.
11. Giorgi, F., & Gutowski, W. J., Jr. (2015). Regional dynamical downscaling and the CORDEX initiative. *Annual Review of Environment and Resources, 40*(1), 467–490.
12. Heede, R. (2014). Tracing anthropogenic carbon dioxide and methane emissions to fossil fuel and cement producers, 1854–2010. *Climatic Change, 122*(1), 229–241.
13. Cox, P. M., Huntingford, C., & Williamson, M. S. (2018). Emergent constraint on equilibrium climate sensitivity from global temperature variability. *Nature, 553*(7688), 319–322.
14. Fowler, H. J., et al. (2021). Towards advancing scientific knowledge of climate change impacts on short-duration rainfall extremes. *Philosophical Transactions of the Royal Society A, 379*(2195), 20190542.
15. Massonnet, F., Fichefet, T., & Goosse, H. (2015). Prospects for improved seasonal Arctic sea ice predictions from multivariate data assimilation. *Ocean Modelling, 88*, 16–25.
16. Carvalho, D. V., Pereira, E. M., & Cardoso, J. S. (2019). Machine learning interpretability: A survey on methods and metrics. *Electronics, 8*(8), 832.
17. Starke, G. M. J. (2022) *Trusting black-box algorithms? Ethical challenges for biomedical machine learning*. University of Basel.
18. Neelin, J. D. (2010). *Climate change and climate modeling*. Cambridge University Press.
19. Van Vuuren, D. P., et al. (2011). The representative concentration pathways: An overview. *Climatic Change, 109*, 5–31.
20. Meinshausen, M., et al. (2020). The shared socio-economic pathway (SSP) greenhouse gas concentrations and their extensions to 2500. *Geoscientific Model Development, 13*(8), 3571–3605.
21. Collins, M. (2015). *Climate modeling* (p. 31). Climate Change and Public Health.
22. Tong, D., et al. (2020). Dynamic projection of anthropogenic emissions in China: Methodology and 2015–2050 emission pathways under a range of socio-economic, climate policy, and pollution control scenarios. *Atmospheric Chemistry and Physics, 20*(9), 5729–5757.
23. Edwards, P. N. (2011). History of climate modeling. *Wiley Interdisciplinary Reviews: Climate Change, 2*(1), 128–139.
24. André, J.-C., et al. (2014). High-performance computing for climate modeling. *Bulletin of the American Meteorological Society, 95*(5), ES97–ES100.
25. Tapiador, F. J., et al. (2020). Regional climate models: 30 years of dynamical downscaling. *Atmospheric Research, 235*, 104785.
26. Perera, A. K. S. S. (2024). *Machine learning and geostatistical approaches for discovery of weather and climate events related to El Niño Phenomena*. Chapman University.
27. Satoh, M., et al. (2019). Global cloud-resolving models. *Current Climate Change Reports, 5*, 172–184.
28. Slater, L. J., et al. (2023). Hybrid forecasting: Blending climate predictions with AI models. *Hydrology and Earth System Sciences, 27*(9), 1865–1889.

29. Gettelman, A., & Rood, R. B. (2016). *Demystifying climate models: A users guide to earth system models*. Springer Nature.
30. Lohmann, G. (2020). Temperatures from energy balance models: The effective heat capacity matters. *Earth System Dynamics, 11*(4), 1195–1208.
31. Soldatenko, S. (2020). Estimating the effect of radiative feedback uncertainties on climate response to changes in the concentration of stratospheric aerosols. *Atmosphere, 11*(6), 654.
32. Chandler, R., Barnes, C., & Brierley, C. (2023). *Decision-relevant characterisation of uncertainty in UK climate projections.*
33. Soldatenko, S., & Colman, R. (2019). Climate variability from annual to multi-decadal timescales in a two-layer stochastic energy balance model: Analytic solutions and implications for general circulation models. *Tellus A: Dynamic Meteorology and Oceanography, 71*(1), 1554421.
34. Tapiador, F. J., et al. (2019). Climate classifications from regional and global climate models: Performances for present climate estimates and expected changes in the future at high spatial resolution. *Atmospheric Research, 228*, 107–121.
35. Jiang, D., Tian, Z., & Lang, X. (2016). Reliability of climate models for China through the IPCC Third to Fifth Assessment Reports. *International Journal of Climatology, 36*(3).
36. Abbas, A., et al. (2022). Evaluation and projection of precipitation in Pakistan using the Coupled Model Intercomparison Project Phase 6 model simulations. *International Journal of Climatology, 13*, 6665–6684.
37. Hewitt, R. J., et al. (2021). Beyond shared socioeconomic pathways (SSPs) and representative concentration pathways (RCPs): Climate policy implementation scenarios for Europe, the US and China. *Climate Policy, 21*(4), 434–454.
38. Bjarke, N. R., et al. (2024). Evaluating large-storm dominance in high-resolution GCMs and observations across the western contiguous United States. *Earth's Future, 12*(6), e2023EF004289.
39. Demory, M.-E., et al. (2020). European daily precipitation according to EURO-CORDEX regional climate models (RCMs) and high-resolution global climate models (GCMs) from the High-Resolution Model Intercomparison Project (HighResMIP). *Geoscientific Model Development, 13*(11), 5485–5506.
40. Chokkavarapu, N., & Mandla, V. R. (2019). Comparative study of GCMs, RCMs, downscaling and hydrological models: A review toward future climate change impact estimation. *SN Applied Sciences, 1*(12), 1698.
41. Alamdari, N., & Hogue, T. S. (2022). Assessing the effects of climate change on urban watersheds: A review and call for future research. *Environmental Reviews, 30*(1), 61–71.
42. Séférian, R., et al. (2019). Evaluation of CNRM earth system model, CNRM-ESM2-1: Role of earth system processes in present-day and future climate. *Journal of Advances in Modeling Earth Systems, 11*(12), 4182–4227.
43. Bonan, G. B., & Doney, S. C. (2018). Climate, ecosystems, and planetary futures: The challenge to predict life in Earth system models. *Science, 359*(6375), eaam8328.
44. Pellerin, F., Porada, P., & Hense, I. (2020). ESD reviews: Evidence of multiple inconsistencies between representations of terrestrial and marine ecosystems in Earth System Models. *Earth System Dynamics Discussions, 2020*, 1–26.
45. Varney, R. M., et al. (2024). Soil carbon-concentration and carbon-climate feedbacks in CMIP6 Earth system models. *Biogeosciences, 21*(11), 2759–2776.
46. Ashok, S., et al. (2022). Developing ecotourism sustainability maximization (ESM) model: A safe minimum standard for climate change mitigation in the Indian Himalayas. *Environmental Monitoring and Assessment, 194*(12), 914.
47. Lee, H., et al. (2022). Development of the UKESM-TOPAZ earth system model (Version 1.0) and preliminary evaluation of its biogeochemical simulations. *Asia-Pacific Journal of Atmospheric Sciences, 2022*, 1–22.
48. Koven, C. D., et al. (2022). Multi-century dynamics of the climate and carbon cycle under both high and net negative emissions scenarios. *Earth System Dynamics, 13*(2), 885–909.

49. Eyring, V., et al. (2024). Pushing the frontiers in climate modelling and analysis with machine learning. *Nature Climate Change*, 1–13.
50. Huntingford, C., et al. (2019). Machine learning and artificial intelligence to aid climate change research and preparedness. *Environmental Research Letters, 14*(12), 124007.
51. Materia, S., et al. (2024). *Artificial intelligence for climate prediction of extremes: State of the art, challenges, and future perspectives* (p. e914). Wiley Interdisciplinary Reviews: Climate Change.
52. Reichstein, M., et al. (2019). Deep learning and process understanding for data-driven Earth system science. *Nature, 566*(7743), 195–204.
53. Gettelman, A., et al. (2016). Components of the climate system. In *Demystifying climate models: A users guide to earth system models* (pp. 13–22).
54. Rind, D., et al. (2020). GISS Model E2. 2: A climate model optimized for the middle atmosphere—Model structure, climatology, variability, and climate sensitivity. *Journal of Geophysical Research: Atmospheres, 125*(10), e2019JD032204.
55. Morrison, H., et al. (2020). Confronting the challenge of modeling cloud and precipitation microphysics. *Journal of Advances in Modeling Earth Systems, 12*(8), e2019MS001689.
56. Ramachandran, S. (2018). Aerosols and climate change: Present understanding, challenges, and future outlook. In: *Land-atmospheric research applications in South and Southeast Asia* (pp. 341–378).
57. Gryspeerdt, E., et al. (2022). Uncertainty in aerosol-cloud radiative forcing is driven by clean conditions. *Atmospheric Chemistry and Physics Discussions, 2022*, 1–13.
58. Liao, H., & Seinfeld, J. H. (1998). Effect of clouds on direct aerosol radiative forcing of climate. *Journal of Geophysical Research: Atmospheres, 103*(D4), 3781–3788.
59. Zhang, B. (2020). The effect of aerosols to climate change and society. *Journal of Geoscience and Environment Protection, 8*(08), 55.
60. Roxy, M., et al. (2024). Future projections for the tropical Indian Ocean. *The Indian ocean and its role in the global climate system* (pp. 469–482). Elsevier.
61. Stocker, T. F. (2013). The ocean as a component of the climate system. *International geophysics* (pp. 3–30). Elsevier.
62. Paynter, D., & Frölicher, T. L. (2015). Sensitivity of radiative forcing, ocean heat uptake, and climate feedback to changes in anthropogenic greenhouse gases and aerosols. *Journal of Geophysical Research: Atmospheres, 120*(19), 9837–9854.
63. Pauthenet, E., et al. (2019). The thermohaline modes of the global ocean. *Journal of Physical Oceanography, 49*(10), 2535–2552.
64. Dutheil, C., et al. (2023). Warming of Baltic Sea water masses since 1850. *Climate Dynamics, 61*(3), 1311–1331.
65. Wang, C., Dong, S., & Munoz, E. (2010). Seawater density variations in the North Atlantic and the Atlantic meridional overturning circulation. *Climate Dynamics, 34*, 953–968.
66. Cai, W., et al. (2020). Climate impacts of the El Niño–southern oscillation on South America. *Nature Reviews Earth & Environment, 1*(4), 215–231.
67. Cai, W., et al. (2021). Changing El Niño-Southern oscillation in a warming climate. *Nature Reviews Earth & Environment, 2*(9), 628–644.
68. Cai, W., et al. (2018). Increased variability of eastern Pacific El Niño under greenhouse warming. *Nature, 564*(7735), 201–206.
69. Wu, P., et al. (2015). The reversibility of CO_2 induced climate change. *Climate Dynamics, 45*, 745–754.
70. Liu, S., et al. (2021). Changes in the global cryosphere and their impacts: A review and new perspective. *Sciences in Cold and Arid Regions, 12*(6), 343–354.
71. Callaghan, T. V., & Johansson, M. (2021). Snow, ice, and the biosphere. *Snow and ice-related hazards, risks, and disasters* (pp. 137–164). Elsevier.
72. Kumar, V., Ranjan, D., & Verma, K. (2021). Global climate change: The loop between cause and impact. *Global climate change* (pp. 187–211). Elsevier.
73. Zhang, L., et al. (2020). Significant methane ebullition from alpine permafrost rivers on the East Qinghai-Tibet Plateau. *Nature Geoscience, 13*(5), 349–354.

74. Mohammed, Z. R. H., et al. (2024). *Global warming and its impact on the climate in 2024.*
75. Ruehr, S., et al. (2023). Evidence and attribution of the enhanced land carbon sink. *Nature Reviews Earth & Environment, 4*(8), 518–534.
76. Vereecken, H., et al. (2022). Soil hydrology in the Earth system. *Nature Reviews Earth & Environment, 3*(9), 573–587.
77. Kobayashi, H., et al. (2024). Assessing transient changes in the ocean carbon cycle during the last deglaciation through carbon isotope modeling. *Climate of the Past, 20*(3), 769–787.
78. Ussiri, D. A., et al. (2017). Introduction to global carbon cycling: an overview of the global carbon cycle. In *Carbon sequestration for climate change mitigation and adaptation* (pp. 61–76).
79. Costa, C., et al. (2021). *Global Warming Potential*(GWP*): Understanding the implications for mitigating methane emissions in agriculture.*
80. Matear, R. J., & Lenton, A. (2018). Carbon–climate feedbacks accelerate ocean acidification. *Biogeosciences, 15*(6), 1721–1732.
81. Rogelj, J., Meinshausen, M., & Knutti, R. (2012). Global warming under old and new scenarios using IPCC climate sensitivity range estimates. *Nature Climate Change, 2*(4), 248–253.
82. Myhre, G., et al. (2013). Aerosols and their relation to global climate and climate sensitivity. *Nature Education Knowledge, 4*(5), 7.
83. Leao, S. Z. (2017). Assessing the trade-off between data quality and spatial resolution for the Thornthwaite Moisture Index mapping. *Journal of Spatial Science, 62*(1), 85–102.
84. Ahmed, M., Ahmad, S., & Kheir, A. M. (2023). Climate change: An overview. In Global agricultural production: Resilience to climate change (pp. 1–30).
85. Bador, M., et al. (2020). Impact of higher spatial atmospheric resolution on precipitation extremes over land in global climate models. *Journal of Geophysical Research: Atmospheres, 125*(13), e2019JD032184.
86. Lucas-Picher, P., et al. (2021). Convection-permitting modeling with regional climate models: Latest developments and next steps. *Wiley Interdisciplinary Reviews: Climate Change, 12*(6), e731.
87. Bustamante, M., et al. (2023). Ten new insights in climate science 2023. *Global Sustainability, 7*, e19.
88. Nisbet, E. G., et al. (2023). Atmospheric methane: Comparison between methane's record in 2006–2022 and during glacial terminations. *Global Biogeochemical Cycles, 37*(8), e2023GB007875.
89. Orihuela Pinto, L. (2022). *Atlantic meridional overturning circulation collapse effects on global climate and feedbacks.* UNSW Sydney.
90. Xie, S.-P. (2022). *Coupled atmosphere-ocean dynamics: From El Niño to climate change.* Elsevier.
91. Rood, R. B. (2019). Validation of climate models: An essential practice. In *Computer simulation validation: Fundamental concepts, methodological frameworks, and philosophical perspectives* (pp. 737–762).
92. Franke, J. G., & Donner, R. V. (2019). Correlating paleoclimate time series: Sources of uncertainty and potential pitfalls. *Quaternary Science Reviews, 212*, 69–79.
93. Salimi, A., Ghobrial, T., & Bonakdari, H. (2024). A comprehensive review of AI-based methods used for forecasting ice jam floods occurrence, severity, timing, and location. *Cold Regions Science and Technology*, 104305.
94. Bochenek, B., & Ustrnul, Z. (2022). Machine learning in weather prediction and climate analyses—Applications and perspectives. *Atmosphere, 13*(2), 180.
95. Zhao, T., et al. (2024). Artificial intelligence for geoscience: Progress, challenges and perspectives. *The Innovation.*
96. Salimi, A., Ghobrial, T., & Bonakdari, H. (2024). How to use artificial intelligence to downscale climate change models' data. *Intelligence systems for earth, environmental and planetary sciences* (pp. 147–183). Elsevier.
97. Egigu, M. (2020). Techniques of filling missing values of daily and monthly rain fall data: a review. *SF Journal of Environmental and Earth Science, 3*(1).

98. Allam, Z., & Dhunny, Z. A. (2019). On big data, artificial intelligence and smart cities. *Cities, 89*, 80–91.
99. Chin, T. M., Vazquez-Cuervo, J., & Armstrong, E. M. (2017). A multi-scale high-resolution analysis of global sea surface temperature. *Remote Sensing of Environment, 200*, 154–169.
100. Pham, B. T., et al. (2020). Development of advanced artificial intelligence models for daily rainfall prediction. *Atmospheric Research, 237*, 104845.
101. Ioannou, K., et al. (2021). Low-cost automatic weather stations in the internet of things. *Information, 12*(4), 146.
102. Lima-Ribeiro, M. S., et al. (2015). EcoClimate: A database of climate data from multiple models for past, present, and future for macroecologists and biogeographers. *Biodiversity Informatics, 10.*
103. Lopatka, A. (2019). Meteorologists predict better weather forecasting with AI. *Physics Today, 72*(5), 32–34.
104. Kim, T., et al. (2022). Near real-time hurricane rainfall forecasting using convolutional neural network models with Integrated Multi-satellitE Retrievals for GPM (IMERG) product. *Atmospheric Research, 270*, 106037.
105. Asadollah, S. B. H. S., et al. (2022). Prediction of heat waves using meteorological variables in diverse regions of Iran with advanced machine learning models. *Stochastic Environmental Research and Risk Assessment* 1–16.
106. Chang, D.-L., et al. (2020). Artificial intelligence methodologies applied to prompt pluvial flood estimation and prediction. *Water, 12*(12), 3552.
107. Mosavi, A., Ozturk, P., & Chau, K.-W. (2018). Flood prediction using machine learning models: Literature review. *Water, 10*(11), 1536.
108. Chao, M. A., et al. (2022). Fusing physics-based and deep learning models for prognostics. *Reliability Engineering & System Safety, 217*, 107961.
109. Cifuentes, J., et al. (2020). Air temperature forecasting using machine learning techniques: A review. *Energies, 13*(16), 4215.
110. Ilyas, A., et al. (2022). *Datamodels: Predicting predictions from training data.* arXiv preprint arXiv:2202.00622
111. Bâra, A., Văduva, A. G., & Oprea, S.-V. (2024). Anomaly detection in weather phenomena: News and numerical data-driven insights into the climate change in Romania's historical regions. *International Journal of Computational Intelligence Systems, 17*(1), 134.
112. Arizmendi, F., & Barreiro, M. (2017). ENSO teleconnections in the southern hemisphere: A climate network view. *Chaos: An Interdisciplinary Journal of Nonlinear Science, 27*(9).
113. Luo, C., et al. (2022). PredRANN: The spatiotemporal attention convolution recurrent neural network for precipitation nowcasting. *Knowledge-Based Systems, 239*, 107900.
114. Wu, Y., & Xue, W. (2024). Data-driven weather forecasting and climate modeling from the perspective of development. *Atmosphere, 15*(6), 689.
115. Wang, Y., et al. (2017). PredRNN: Recurrent neural networks for predictive learning using spatiotemporal LSTMS. In *Advances in neural information processing systems*(p. 30).
116. Liu, J., et al. (2018). TD-LSTM: Temporal dependence-based LSTM networks for marine temperature prediction. *Sensors, 18*(11), 3797.
117. Alsumaiei, A. A., & Alrashidi, M. S. (2020). Hydrometeorological drought forecasting in hyper-arid climates using nonlinear autoregressive neural networks. *Water, 12*(9), 2611.
118. Martel, J.-L., et al. (2024). Assessing the adequacy of traditional hydrological models for climate change impact studies: A case for long-short-term memory (LSTM) neural networks. *EGUsphere, 2024*, 1–44.
119. Kadow, C., Hall, D. M., & Ulbrich, U. (2020). Artificial intelligence reconstructs missing climate information. *Nature Geoscience, 13*(6), 408–413.
120. de Melo, C. M., et al. (2022). Next-generation deep learning based on simulators and synthetic data. *Trends in Cognitive Sciences, 26*(2), 174–187.
121. Verma, Y., Heinonen, M., & Garg, V. (2024). Climode: Climate and weather forecasting with physics-informed neural odes. arXiv preprint arXiv:2404.10024

122. Mienye, I. D., & Sun, Y. (2022). A survey of ensemble learning: Concepts, algorithms, applications, and prospects. *IEEE Access, 10*, 99129–99149.
123. Hassani, H., Huang, X., & Silva, E. (2019). Big data and climate change. *Big Data and Cognitive Computing, 3*(1), 12.
124. Schnase, J. L. (2016). Climate analytics as a service. *Cloud computing in ocean and atmospheric sciences* (pp. 187–219). Elsevier.
125. Rodrigues, E. R., et al. (2018). DeepDownscale: A deep learning strategy for high-resolution weather forecast. In *2018 IEEE 14th International Conference on e-Science (e-Science)*. IEEE.
126. Castelvecchi, D. (2016). Can we open the black box of AI? *Nature News, 538*(7623), 20.
127. Baghanam, A. H., et al. (2024). A systematic review of predictor screening methods for downscaling of numerical climate models. *Earth-Science Reviews, 104773.*
128. Quesada-Chacón, D., Barfus, K., & Bernhofer, C. (2022). Repeatable high-resolution statistical downscaling through deep learning. *Geoscientific Model Development, 15*(19), 7353–7370.
129. Balmaceda-Huarte, R., et al. (2024). On the use of convolutional neural networks for downscaling daily temperatures over southern South America in a climate change scenario. *Climate Dynamics, 62*(1), 383–397.
130. Sato, H., & Ise, T. (2022). Predicting global terrestrial biomes with the LeNet convolutional neural network. *Geoscientific Model Development, 15*(7), 3121–3132.
131. Rosenzweig, C., & Hillel, D. (2008). *Climate variability and the global harvest: Impacts of El Niño and other oscillations on agro-ecosystems*. Oxford University Press.
132. Timmermann, A., et al. (2018). El Niño–southern oscillation complexity. *Nature, 559*(7715), 535–545.
133. Wang, G.-G., et al. (2023). ENSO analysis and prediction using deep learning: A review. *Neurocomputing, 520*, 216–229.
134. Nobre, G. G., et al. (2019). Achieving the reduction of disaster risk by better predicting impacts of El Niño and La Niña. *Progress in Disaster Science, 2*, 100022.
135. Li, Y., Strapasson, A., & Rojas, O. (2020). Assessment of El Niño and La Niña impacts on China: Enhancing the early warning system on food and agriculture. *Weather and Climate Extremes, 27*, 100208.
136. Zheng, G., et al. (2020). Purely satellite data-driven deep learning forecast of complicated tropical instability waves. *Science Advances, 6*(29), eaba1482.
137. Young, C.-C., et al. (2024). Accurate reconstruction of satellite-derived SST under cloud and cloud-free areas using a physically-informed machine learning approach. *Remote Sensing of Environment, 313*, 114339.
138. Bertrand, A., et al. (2020). *El Niño Southern Oscillation (ENSO) effects on fisheries and aquaculture* (Vol. 660). Food & Agriculture Org.
139. Kopp, R. E., et al. (2019). Usable science for managing the risks of sea-level rise. *Earth's Future, 7*(12), 1235–1269.
140. Climate Change Initiative Coastal Sea Level. (2020). Coastal sea level anomalies and associated trends from Jason satellite altimetry over 2002–2018. *Scientific Data, 7*(1), 357.
141. Ayinde, A. S., Yu, H., & Wu, K. (2023). Sea level variability and modeling in the Gulf of Guinea using supervised machine learning. *Scientific Reports, 13*(1), 21318.
142. Pang, R., et al. (2023). An asynchronous parallel I/O framework for mass conservation ocean model. *Applied Sciences, 13*(24), 13230.
143. Bahari, N. A. A. B. S., et al. (2023). Predicting sea level rise using artificial intelligence: a review. *Archives of Computational Methods in Engineering, 30*(7), 4045–4062.
144. Albahri, A., et al. (2024). A systematic review of trustworthy artificial intelligence applications in natural disasters. *Computers and Electrical Engineering, 118*, 109409.
145. Farahmand, H., Xu, Y., & Mostafavi, A. (2023). A spatial–temporal graph deep learning model for urban flood nowcasting leveraging heterogeneous community features. *Scientific Reports, 13*(1), 6768.
146. Tiggeloven, T., et al. (2021). Exploring deep learning capabilities for surge predictions in coastal areas. *Scientific Reports, 11*(1), 17224.

147. Wei, C.-C., & Chang, H.-C. (2021). Forecasting of typhoon-induced wind-wave by using convolutional deep learning on fused data of remote sensing and ground measurements. *Sensors, 21*(15), 5234.
148. Chen, R., Zhang, W., & Wang, X. (2020). Machine learning in tropical cyclone forecast modeling: A review. *Atmosphere, 11*(7), 676.
149. Giffard-Roisin, S., et al. (2020). Tropical cyclone track forecasting using fused deep learning from aligned reanalysis data. *Frontiers in Big Data, 3*, 1.
150. Wu, H., & Levinson, D. (2021). The ensemble approach to forecasting: A review and synthesis. *Transportation Research Part C: Emerging Technologies, 132*, 103357.
151. Nishant, R., Kennedy, M., & Corbett, J. (2020). Artificial intelligence for sustainability: Challenges, opportunities, and a research agenda. *International Journal of Information Management, 53*, 102104.
152. Adegbite, A. O., et al. (2024). Artificial Intelligence in climate change mitigation: A review of predictive modeling and data-driven solutions for reducing greenhouse gas emissions.
153. Khallaf, A. N. (2024). Using Ai to help reduce the effect of global warming. *Power System Technology, 48*(1), 1927–1947.
154. Schwartz, R., et al. (2022). *Towards a standard for identifying and managing bias in artificial intelligence* (Vol. 3.). US Department of Commerce, National Institute of Standards and Technology.
155. Ghaffarian, S., Taghikhah, F. R., & Maier, H. R. (2023). Explainable artificial intelligence in disaster risk management: Achievements and prospective futures. *International Journal of Disaster Risk Reduction, 98*, 104123.
156. Minh, D., et al. (2022). Explainable artificial intelligence: A comprehensive review. *Artificial Intelligence Review* 1–66.
157. Lauterbach, A. (2019). Artificial intelligence and policy: Quo vadis? *Digital Policy, Regulation and Governance, 21*(3), 238–263.
158. Jain, H., et al. (2023). AI-enabled strategies for climate change adaptation: Protecting communities, infrastructure, and businesses from the impacts of climate change. *Computational Urban Science, 3*(1), 25.
159. Turner, B. L., et al. (2016). System dynamics modeling for agricultural and natural resource management issues: Review of some past cases and forecasting future roles. *Resources, 5*(4), 40.
160. Şen, Z. (2015). *Applied drought modeling, prediction, and mitigation.* Elsevier.
161. Drogkoula, M., Kokkinos, K., & Samaras, N. (2023). A comprehensive survey of machine learning methodologies with emphasis in water resources management. *Applied Sciences, 13*(22), 12147.
162. Allam, Z. (2020). *The rise of autonomous smart cities: Technology, economic performance and climate resilience.* Springer Nature.
163. Singh, S., & Goyal, M. K. (2023). Enhancing climate resilience in businesses: The role of artificial intelligence. *Journal of Cleaner Production, 418*, 138228.
164. Amiri, Z., Heidari, A., & Navimipour, N. J. (2024). Comprehensive survey of artificial intelligence techniques and strategies for climate change mitigation. *Energy* 132827.
165. Qerimi, Q., & Sergi, B. S. (2022). The case for global regulation of carbon capture and storage and artificial intelligence for climate change. *International Journal of Greenhouse Gas Control, 120*, 103757.
166. Soltani, K., et al. (2024). Advanced forecasting of drought zones in Canada using deep learning and CMIP6 projections. *Climate, 12*(8), 119.
167. Ebtehaj, I., Bonakdari, H., & Gharabaghi, B. (2023). AI-enhanced flood forecasting: Harnessing upstream data for downstream protection. *Journal of Applied Research in Water and Wastewater, 10*(2), 119–132.
168. Varsha, V., Naganandini, S., & Hariharan, C. (2024). Utilizing AI and machine learning for natural disaster management: Predicting natural disasters with AI and machine learning. In *Internet of Things and AI for natural disaster management and prediction* (pp. 279–304). IGI Global.

169. Mana, A., et al. (2024). Sustainable AI-based production agriculture: Exploring AI applications and implications in agricultural practices. In *Smart agricultural technology* (p. 100416).
170. Ramampiandra, E. C., et al. (2023). A comparison of machine learning and statistical species distribution models: Quantifying overfitting supports model interpretation. *Ecological Modelling, 481*, 110353.
171. Payne, M. R., et al. (2016). Uncertainties in projecting climate-change impacts in marine ecosystems. *ICES Journal of Marine Science, 73*(5), 1272–1282.
172. Khosravi, F., Izbirak, G., & Shavarani, S. M. (2021). Application of bootstrap re-sampling method in statistical measurement of sustainability. *Socio-Economic Planning Sciences, 75*, 100781.
173. Gerges, F., et al. (2022). Bayesian multi-head convolutional neural networks with Bahdanau attention for forecasting daily precipitation in climate change monitoring. In *Joint European Conference on Machine Learning and Knowledge Discovery in Databases*. Springer.
174. Donnelly, J., Daneshkhah, A., & Abolfathi, S. (2024). Forecasting global climate drivers using Gaussian processes and convolutional autoencoders. *Engineering Applications of Artificial Intelligence, 128*, 107536.

Chapter 5
Revolutionizing Natural Resource Management with Artificial Intelligence: A Review

Balendra V. S. Chauhan, Ajitanshu Vedrtnam, Kevin P. Wyche, and Sneha Verma

Abstract This chapter provides a comprehensive review of the applications of Artificial Intelligence (AI) in Natural Resource Management (NRM). It explores how AI, combined with technologies like remote sensing, the Internet of Things (IoT), and big data analytics, is transforming traditional approaches to NRM by enhancing data collection, processing, and interpretation capabilities. AI algorithms, particularly machine learning and deep learning models, demonstrate remarkable accuracy in analyzing complex datasets, extracting valuable insights, and making precise predictions, thereby facilitating informed and timely decision-making. The chapter examines various applications of AI in NRM, including water resource management, forest management, mineral and mining management, and biodiversity conservation. Specific examples include the use of Convolutional Neural Networks (CNNs) for land cover classification and deforestation monitoring, Long Short-Term Memory (LSTM) networks for streamflow prediction and flood forecasting, and AI-powered object detection algorithms for monitoring wildlife populations and detecting illegal activities. The chapter also discusses the benefits of AI in NRM, such as improved efficiency and accuracy through automation, enhanced data utilization through the

A. Vedrtnam
Mineral and Energy Economy Institute, Polish Academy of Sciences, Kraków, Poland
e-mail: ajitanshu.m@invertis.org

B. V. S. Chauhan · K. P. Wyche (✉)
Centre for Earth Observation Science, University of Brighton, Brighton, UK
e-mail: kevinpwyche@googlemail.com

B. V. S. Chauhan
e-mail: b.chauhan@brighton.ac.uk

A. Vedrtnam
Institute of Construction and Architecture, Slovak Academy of Science, Bratislava, Slovakia

Department of Mechanical Engineering, Invertis University, Bareilly, India

B. V. S. Chauhan · S. Verma
School of Engineering, Newcastle University, Newcastle Upon Tyne, United Kingdom
e-mail: Sneha.verma@newcastle.ac.uk

A. Vedrtnam et al., *Prospects of Artificial Intelligence in the Environment*,
https://doi.org/10.1007/978-981-96-6863-2_5

integration of diverse data sources, cost savings through optimized resource allocation, and increased accessibility to advanced analytical tools for a wider range of stakeholders. However, the chapter acknowledges the challenges and considerations associated with AI implementation, including data quality and availability, ethical and privacy concerns, integration with existing systems, and skill requirements. Addressing these challenges requires collaborative efforts among scientists, AI researchers, policymakers, and stakeholders to develop robust, interpretable, and ethically aligned AI systems for environmental applications. The chapter concludes by emphasizing the transformative potential of AI in revolutionizing NRM and ensuring the responsible and sustainable use of Earth's resources.

Keywords Artificial Intelligence (AI) · Natural Resource Management (NRM) · Environmental monitoring · Internet of Things (IoT) · Prediction · Ecosystem management · Natural disaster management · Renewable energy optimization · Waste management · Conservation · Wildlife protection · Water resource management · Forest management · Mineral and mining management · Biodiversity conservation · Convolutional Neural Networks (CNNs) and Long Short-Term Memory (LSTM)

1 Introduction

Natural resources, including water, minerals, forests, and biodiversity, are critical to sustaining human life and economic activities. Efficient management of these resources is essential for ensuring their availability for future generations while also mitigating environmental impacts. In recent years, AI has emerged as a transformative tool in natural resource management, providing advanced capabilities for monitoring, analysis, and decision-making. This chapter explores the applications, benefits, and challenges of AI in managing natural resources, offering insights into how AI technologies are reshaping this field. The integration of AI in natural resource management has revolutionized traditional approaches by enhancing data collection, processing, and interpretation capabilities. Remote sensing technologies, coupled with AI-driven image classification and object detection algorithms, have significantly improved our ability to monitor and assess large-scale environmental changes. For instance, Convolutional Neural Networks (CNNs) have demonstrated remarkable accuracy in classifying land cover types and detecting specific features in satellite imagery, enabling more precise tracking of deforestation, urban expansion, and changes in water bodies [1]. AI-powered predictive modeling and simulation have also transformed our understanding of complex ecological systems. Machine learning algorithms, such as Random Forests and Support Vector Machines, are being employed to model ecosystem dynamics, predict species distributions, and assess the impacts of climate change on natural resources. These models integrate diverse data sources, including remote sensing data, climate projections, and field observations, to provide comprehensive insights into ecosystem behavior and inform adaptive management strategies [2]. In the realm of water resource management, AI techniques like Long Short-Term Memory (LSTM) networks have enhanced the accuracy of streamflow predictions and flood forecasting. By capturing complex non-linear relationships in hydrological systems, these models enable more reliable water resource management strategies, crucial for addressing water scarcity and flood risks in a changing climate [3]. The application of AI in mineral exploration and resource estimation has led to more efficient and targeted exploration strategies. Machine learning algorithms analyze geological, geochemical, and geophysical data to identify potential areas of mineralization with higher accuracy than traditional methods. This not only reduces the environmental impact of exploration activities but also improves the economic viability of mining operations [4]. Despite these advancements, the implementation of AI in natural resource management faces several challenges. These include the need for high-quality, comprehensive datasets; the complexity of interpreting AI model outputs for decision-making; and ethical considerations surrounding data privacy and the potential for AI to exacerbate existing inequalities in resource access. Addressing these challenges requires interdisciplinary collaboration between environmental scientists, AI researchers, policymakers, and stakeholders to develop robust, interpretable, and ethically aligned AI systems for environmental applications.

1.1 AI Technologies in Natural Resource Management

AI technologies are transforming the management of natural resources through advanced data analysis and predictive capabilities. This section delves into how AI is leveraged in remote sensing, predictive modeling, and automated monitoring, emphasizing their roles and advancements. Figure 1 shows the comprehensive flowchart of how AI Technologies are being used in natural resource management.

1.1.1 Remote Sensing and Geospatial Analysis

Remote sensing technologies, such as satellites and drones, provide vast amounts of geospatial data that are pivotal for effective natural resource management. The integration of AI with remote sensing enables more sophisticated analysis and interpretation of this data.

- Image Classification: AI-driven image classification employs machine learning algorithms to categorize and analyze satellite and drone imagery. Convolutional Neural Networks are particularly effective in this domain, thanks to their ability to detect spatial hierarchies in images [5]. CNNs can identify various land cover types, such as forests, urban areas, and water bodies, with high accuracy. This capability is crucial for monitoring land use changes, assessing deforestation rates, and tracking urban expansion over time. Advanced models, including ResNet and

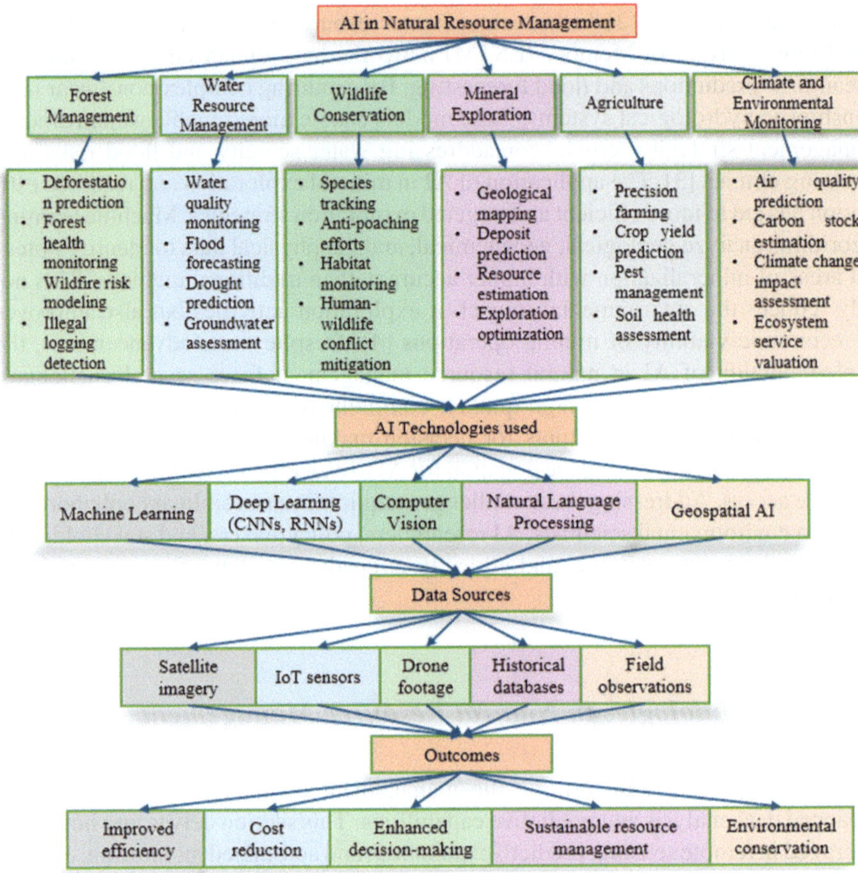

Fig. 1 Comprehensive overview of how AI is applied across various aspects of natural resource management, the technologies used, data sources, and the potential outcomes

U-Net, have further enhanced classification precision by addressing issues like image resolution and variability in land cover types [6].

- Object Detection: Object detection algorithms powered by AI are designed to identify and locate specific features within high-resolution images. These algorithms, including YOLO (You Only Look Once) and Faster R-CNN, are instrumental in detecting illegal activities such as logging or poaching, as well as monitoring wildlife populations [7, 8]. By pinpointing and tracking objects of interest, AI enables timely intervention and enforcement actions, improving resource protection efforts. The integration of object detection with spatial data analysis enhances the ability to manage and conserve natural resources effectively.
- Change Detection: AI also facilitates change detection by comparing geospatial data from different time periods. Techniques such as differencing and post-classification comparison, combined with AI, can identify changes in land cover,

vegetation health, and environmental conditions [9, 10]. This capability supports proactive management strategies and helps in understanding the impact of human activities and natural events on ecosystems.

1.1.2 Predictive Modeling and Simulation

Predictive modeling using AI enhances the ability to forecast natural resource availability and environmental changes, providing valuable insights for planning and management.

- Climate Modeling: Oliver [11] stated that AI techniques, including neural networks and ensemble methods, significantly improve climate models by increasing the accuracy and resolution of predictions. Machine learning models, such as Recurrent Neural Networks (RNNs) and LSTMs, are employed to analyze historical climate data and simulate future climate scenarios [12]. These models enhance the understanding of complex climate systems and improve predictions related to temperature variations, precipitation patterns, and extreme weather events. The incorporation of AI helps in addressing limitations of traditional climate models and provides more reliable forecasts.
- Resource Demand Forecasting: AI models are used to predict future resource demands by analyzing historical data and current trends. Time-series forecasting methods, such as ARIMA (Auto Regressive Integrated Moving Average) and Prophet, are commonly employed to project future demands for resources like water, energy, and minerals [13]. Machine learning algorithms, including regression models and clustering techniques, further refine these forecasts by identifying patterns and anomalies in resource consumption data [14]. Accurate demand forecasting aids in resource planning, allocation, and sustainability efforts.
- Ecosystem Simulation: AI-driven simulation models, such as agent-based models and system dynamics models, simulate the interactions within ecosystems and predict their responses to various management scenarios. These simulations help in understanding the potential impacts of different policies and interventions on ecosystem health and resilience.

1.1.3 Automated Monitoring and Surveillance

AI enhances the monitoring and surveillance of natural resources through automation, enabling real-time data collection and analysis.

- Wildlife Monitoring: AI-powered monitoring systems utilize camera traps, acoustic sensors, and other technologies to track wildlife populations and behaviors [15]. Machine learning algorithms analyze images and sounds to identify species, monitor movements, and assess habitat use. These automated systems provide valuable data on animal populations, migration patterns, and ecological interactions, contributing to more effective conservation strategies. Techniques

such as automated species recognition and behavior classification further enhance the capabilities of wildlife monitoring systems [16].
- Pollution Detection: AI algorithms analyze data from environmental sensors to detect and quantify pollutants in air, water, and soil [17–20]. Techniques such as anomaly detection and pattern recognition are employed to identify pollution sources and assess their impact [21]. AI-enhanced pollution monitoring systems provide real-time insights into environmental quality, enabling more effective pollution control measures and regulatory compliance. The integration of AI with sensor networks improves the sensitivity and accuracy of pollution detection, leading to better environmental management.
- Disaster Response: AI also plays a role in disaster response by analyzing data from various sources, including satellite imagery and social media, to assess the extent of damage and coordinate relief efforts [22]. Machine learning models help in mapping affected areas, predicting the spread of disasters, and optimizing resource distribution during emergencies.

2 Applications of AI in Natural Resource Management

AI has diverse applications across various domains of natural resource management, providing innovative solutions for monitoring, optimization, and conservation. This section elaborates on the key applications of AI in water resource management, forest management, mineral and mining management, and biodiversity conservation.

2.1 Water Resource Management

The integration of AI with Internet of Things (IoT) technologies has revolutionized water resource management. Employing AI allows for advanced data collection, analysis, and interpretation, enabling proactive management of water resources. The data-driven approach offers advantages over conventional techniques such as real-time monitoring, high-resolution data, and predictive modeling.

Data-Driven Water Quality Assessment

Data-driven water quality assessment is crucial for addressing challenges like increasing water scarcity, pollution from various sources, and the impact of climate change. This approach involves using AI techniques to analyze water quality data and identify potential issues. For example, data analytics can be used to monitor the concentrations of contaminants like microorganisms, heavy metals, and chemicals in water, pinpointing regions where there is a chance of poor water quality and enabling preventative measures.

AI Techniques for Water Quality Monitoring

AI algorithms, including machine learning techniques such as Support Vector Machines (SVMs), Deep Neural Networks (DNNs), and K-nearest neighbors, have been successfully applied to assess water quality [23]. These techniques offer increased accuracy in measuring groundwater quality compared to traditional methods. A study found that an IoT system achieved 95% accuracy in measuring pH, turbidity, TDS, and temperature, compared to only 85% accuracy using traditional methods [23]. Various machine learning techniques have been implemented for assessing water quality. Table 1 summarizes the tasks, algorithms, sample sizes, input parameters, and evaluation results from several studies:

Demand Forecasting: AI enhances water demand forecasting through predictive analytics. Machine learning models, such as time-series forecasting and regression techniques, analyze historical consumption data, climate conditions, and demographic trends to predict future water needs. This information is crucial for urban planners and agricultural managers to optimize water distribution, allocate resources efficiently, and implement conservation strategies. AI also helps in managing water resources during droughts and peak usage periods by adjusting supply strategies based on forecasted demand.

Flood Prediction and Management: AI improves flood prediction and management by integrating various data sources, including weather forecasts, river flow measurements, and historical flood records. Machine learning models, such as ensemble methods and deep learning algorithms, analyze these datasets to predict flood events and assess their potential impacts. AI-driven flood forecasting systems provide accurate and timely warnings, helping communities prepare for and respond to flood events more effectively. Additionally, AI supports flood management by optimizing response strategies and resource allocation during emergencies.

2.2 *Forest Management*

AI technologies are revolutionizing forest management by enhancing monitoring and decision-making processes.

Deforestation Detection: AI algorithms analyze satellite imagery and aerial data to detect and quantify deforestation activities [10]. Techniques such as change detection algorithms and supervised classification models are used to identify areas of forest loss and assess the extent of deforestation. AI helps in tracking illegal logging activities, monitoring land use changes, and supporting conservation efforts. The process to track illegal logging activities is shown in Fig. 2. By providing detailed and up-to-date information, AI enables more effective policy enforcement and sustainable forest management practices.

Table 1 Studies showing how AI has been used in water quality monitoring and management

Prediction target	Machine learning model	Key input parameters	Performance evaluation	Reference
Water Quality Index (WQI)	Random Forest, Gradient Boosting	Dissolved oxygen, pH, conductivity, biological oxygen demand, nitrate, fecal coliform, total coliform	GB achieved 99.5% accuracy for WQC, MLP regressor achieved R^2 of 99.8% for WQI	[24]
Water Quality Index (WQI), Water Quality Classification (WQC)	Adaptive Neuro-Fuzzy Inference System (ANFIS); Feed-forward Neural Network (FFNN); K-Nearest Neighbors (KNN)	DO, pH, conductivity, biological oxygen demand, nitrate, fecal coliform, temp, total coliform	ANFIS accurately predicted WQI with a regression coefficient of 96.17% during testing. FFNN achieved the highest accuracy (100%) for WQC. KNN achieved 80.63% accuracy for WQC	[25]
Water Quality Condition (WQC)	Generalised linear model (GLM) and boosted regression tree (BRT) model	River discharge (Q)	The study can predict an increase in annual concentration nitrate, suspended sediments and phosphorus by 16–63%, 45–146% and 38–100%, respectively, in 2025–2034	[26]
Biological oxygen demand (BOD) prediction	Deep neural network (DNN), Support vector regression (SVR), Random Forest (RF)	Geospatial and temporal data Site-specific physical conditions Atmospheric factors Water quality indicators (DO, temperature, salinity) Biological markers Optical properties Chemical pollutants Nutrient levels pH and conductivity Water clarity and density measurements	Deep Neural Networks (DNNs) outperform traditional models, reducing Root Mean Square Error (RMSE) by 19.20–25.16%	[27]

(continued)

Table 1 (continued)

Prediction target	Machine learning model	Key input parameters	Performance evaluation	Reference
River Water Temperature (RWT)	Support Vector Regression (SVR)	Maximum air temperature	$R2 = 0.84$, $KGE = 0.86$, $MSE = 0.99$, $RMSE = 0.99$, $RSR = 0.40$, $NSE = 0.84$, $MAE = 0.77$ (daily); $R2 = 0.88$, $KGE = 0.88$, $MSE = 0.61$, $RMSE = 0.78$, $RSR = 0.39$, $NSE = 0.88$, $MAE = 0.57$ (monthly)	[28]
River water quality level	Deep Neural Network (DNN) with Long Short-Term Memory (LSTM)	pH, Dissolved Oxygen (DO), Biochemical Oxygen Demand (BOD), Fecal Coliform (FC), Nitrates (NO_3), Turbidity	Over 90% accuracy, even with 20% noisy labels. Precision, recall, and F1-score are also reported	[29]
Water quality (DO and CODMn)	Savitzky-Golay filter and Long Short-Term Memory (LSTM) based Encoder-Decoder Neural Network (SE-LSTM)	Time series data of water quality	Outperforms benchmark models (ARIMA, ANN, SVR, XGBoost, LSTM) on RMSE, MAE, and MAPE for multi-step prediction of DO and CODMn	[30]

Forest Health Monitoring: AI models assess forest health by analyzing data related to tree growth, pest infestations, and disease outbreaks. Remote sensing data, combined with machine learning techniques such as clustering and anomaly detection, is used to identify patterns and trends in forest health. This information is crucial for managing forest ecosystems, detecting early signs of pest invasions or diseases, and implementing targeted management actions to maintain forest biodiversity and productivity.

Fire Risk Assessment: AI also plays a role in assessing and managing wildfire risks. Figure 3 shows how wildfires can be predicted using machine learning. Machine learning models analyze weather data, vegetation conditions, and historical fire records to predict fire risks and identify vulnerable areas [31]. AI-driven fire risk assessment systems help in deploying preventive measures and resources more effectively, reducing the impact of wildfires on ecosystems and communities.

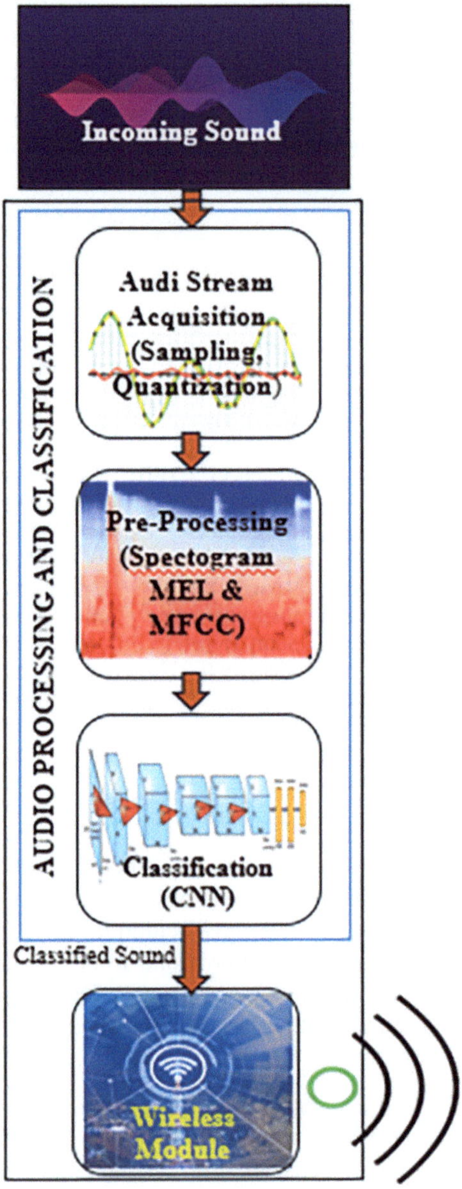

Fig. 2 System for monitoring illegal tree logging

2.3 *Mineral and Mining Management*

AI applications in the mineral and mining sectors enhance exploration, extraction, and operational efficiency.

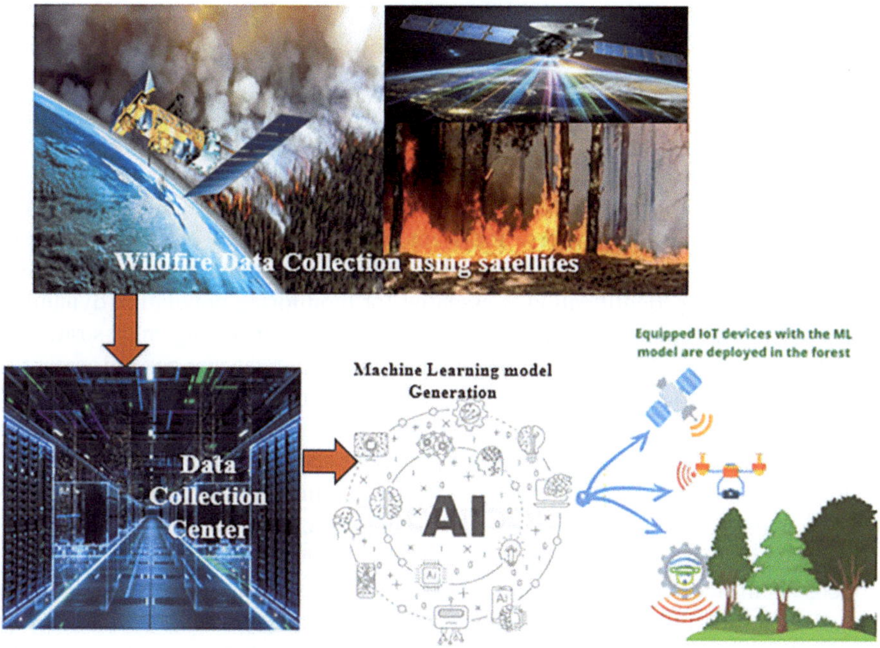

Fig. 3 A comprehensive IoT and ML model process designed for the early detection and monitoring of forest fires, to promptly identify and respond to potential wildfire threats

- Exploration: AI algorithms analyze geological and geospatial data to identify potential mineral deposits [32]. Machine learning techniques, such as supervised learning and pattern recognition, are used to process complex datasets, including geophysical surveys and mineralogical data [33]. AI-driven exploration tools improve the accuracy of mineral deposit predictions, reduce exploration costs, and increase discovery rates by identifying promising areas with higher precision [34].
- Resource Extraction: AI-driven automation and robotics enhance efficiency in mining operations. Automated systems, including autonomous drilling rigs and robotic ore handling, improve the precision and speed of resource extraction [35]. Machine learning models optimize mining processes by analyzing data from equipment sensors, production metrics, and operational conditions. AI also supports waste management by predicting and minimizing waste generation during extraction and processing activities.
- Process Optimization: AI enhances the efficiency of mineral processing by optimizing parameters and workflows [36]. Techniques such as reinforcement learning and optimization algorithms are used to improve the separation and concentration of valuable minerals, leading to higher recovery rates and reduced processing costs.

2.4 Biodiversity Conservation

AI plays a significant role in supporting biodiversity conservation efforts through advanced monitoring and modeling techniques.

- Species Identification: AI-powered image recognition tools, including Convolutional neural networks (CNNs), help in identifying and cataloging species from photographs and videos. These tools automate the process of species identification, contributing to the creation and maintenance of biodiversity databases. Accurate species identification is essential for monitoring population dynamics, assessing conservation status, and implementing effective management strategies.
- Habitat Modeling: Machine learning models predict habitat changes and assess the impacts of environmental changes on species distribution and ecosystem services. Techniques such as habitat suitability modeling and species distribution modeling use environmental variables, climate data, and species observations to predict potential changes in habitat availability and quality. AI-driven habitat models support conservation planning by identifying critical habitats and assessing the effects of climate change and land use changes on biodiversity.
- Ecosystem Services Assessment: AI also aids in assessing ecosystem services, such as pollination, water purification, and carbon sequestration. Machine learning models analyze environmental data to quantify the contributions of ecosystems to human well-being and identify areas where conservation efforts can enhance ecosystem service delivery.

Monitoring target	Data source	Data analysis/ method	Potential benefits/ challenges	References
Fish abundance, behavior, and distribution	Underwater video footage	Deep learning algorithms	**Benefits**: Automates analysis of fish populations, potentially faster, more accurate, and consistent than manual analysis. **Challenges**: Need for large, labeled datasets to train deep learning models. Potential for biases in model predictions if training data is not representative of the target environment. Uncertainty in model outputs and their impact on ecological inferences needs further investigation	[37, 38]

(continued)

(continued)

Monitoring target	Data source	Data analysis/ method	Potential benefits/ challenges	References
Deforestation prediction and monitoring	Satellite imagery, sensor data (e.g., chainsaw sounds)	AI, machine learning (tensor flow), spatial modeling	**Benefits**: Real-time deforestation alerts, identification of deforestation drivers, proactive land-use decision support. **Challenges**: Data access and quality, model accuracy and reliability, infrastructure limitations in remote areas, potential ethical concerns regarding surveillance and community rights	[39]
Forest inventory, mapping, carbon, and biomass estimation	Satellite imagery, LiDAR, Drone imagery, Field survey data	AI, machine learning, cloud computing, GIS	**Benefits**: Automated and accurate forest resource assessment, improved carbon stock estimation for REDD + initiatives, efficient and cost-effective data collection. **Challenges**: Availability and accessibility of high-resolution data, expertise in AI and remote sensing, computational resources and infrastructure	[40–43]
Automated reforestation and afforestation	Satellite imagery, Drone imagery, GPS, seed dispersal technologies	AI, machine learning, drone swarms, robotics	**Benefits**: Accelerated reforestation efforts, increased planting efficiency and coverage, monitoring of reforestation success. **Challenges**: Initial investment costs, technological expertise, potential ecological impacts of large-scale interventions	[44]

(continued)

(continued)

Monitoring target	Data source	Data analysis/ method	Potential benefits/ challenges	References
Hazard assessment and prediction (e.g., wildfires, pests, diseases)	Sensor data, satellite imagery, climate and weather data, historical forest data	AI, machine learning, predictive modeling	**Benefits**: Early warning systems for forest threats, improved forest health monitoring, informed decision-making for risk mitigation. **Challenges**: Data integration and analysis, model accuracy and generalization, real-time data processing and dissemination	[45–49]
Wildlife poaching and illegal logging detection	Sensor networks, acoustic monitoring, drone surveillance, camera traps	AI, machine learning, pattern recognition, real-time alerts	**Benefits**: Increased detection rates, deterrence of illegal activities, improved enforcement and protection of wildlife and forests. **Challenges**: Cost and maintenance of technology, potential for false positives, ethical considerations related to surveillance and privacy	[50–53]
Biodiversity conservation	Remote sensing data (landsat and sentinel), citizen science platforms (iNaturalist, eBird), Global Biodiversity Databases (GBIF, NatureServe)	Big data analytics, data integration, predictive modeling	**Benefits**: Comprehensive monitoring of biodiversity trends, improved understanding of species distributions, enhanced conservation planning, engagement of citizen scientists. **Challenges**: Data quality and standardization, data accessibility and sharing, computational resources for large-scale analysis	[54, 55]
Species distribution and habitat suitability	Remote sensing data, citizen science platforms (iNaturalist, eBird), global biodiversity databases (GBIF, NatureServe)	Big data analytics, data integration, predictive modeling	**Benefits**: Helps to understand and predict species distribution and habitat suitability for conservation planning. **Challenges**: Data quality and standardization, data accessibility and sharing, need for computational resources for large-scale analysis	[56]

3 Benefits of AI in Natural Resource Management

AI technologies offer significant advantages in natural resource management by improving efficiency, accuracy, and accessibility. Intelligent systems, including AI, are playing an increasingly important role in natural resource management by providing a wide array of tools and techniques for monitoring, analyzing, and managing these resources more effectively and sustainably [57, 58]. AI technologies, such as machine learning and predictive modeling, can analyze large amounts of data from various sources like remote sensing, sensor networks, and citizen science platforms, enabling researchers and managers to better understand patterns, trends, and potential risks to natural resources [59, 60]. For example, AI can be used to monitor wildlife populations, detect illegal poaching activities, map critical habitats, and predict species distribution and habitat suitability [61]. AI-powered decision support systems can help to optimize resource allocation, such as determining efficient irrigation schedules or planning sustainable land use practices [62]. Furthermore, intelligent systems facilitate public participation in resource management through information sharing on social media, leading to more effective strategies and the emergence of positive initiatives [63]. These advancements empower stakeholders to make informed decisions, improve resource allocation, predict and mitigate environmental risks, and ultimately contribute to the sustainable management of our planet's resources.

This section explores the key benefits of AI in this field.

3.1 Improved Efficiency and Accuracy

AI technologies greatly enhance the efficiency and accuracy of natural resource management practices through automation and advanced data analysis.

- Automated Data Analysis: AI systems automate the processing and analysis of large volumes of data, such as satellite imagery, sensor data, and environmental records [64, 65]. Machine learning algorithms, including deep learning models and pattern recognition techniques, can quickly analyze complex datasets to extract actionable insights. This automation reduces the time and effort required for data processing, allowing for more rapid decision-making. Figure 4 shows the applications of AI in the natural resource management.
- Enhanced Decision-Making: AI improves decision-making by providing accurate and timely information. Predictive models can forecast future resource availability and demand with high precision, enabling managers to make well-informed decisions. AI-driven tools also support real-time monitoring and analysis, helping to identify issues and implement interventions more effectively.
- Reduction of Human Error: By leveraging AI for data analysis and interpretation, the potential for human error is minimized [66]. AI algorithms provide consistent and objective analyses, reducing biases and inaccuracies that may arise from

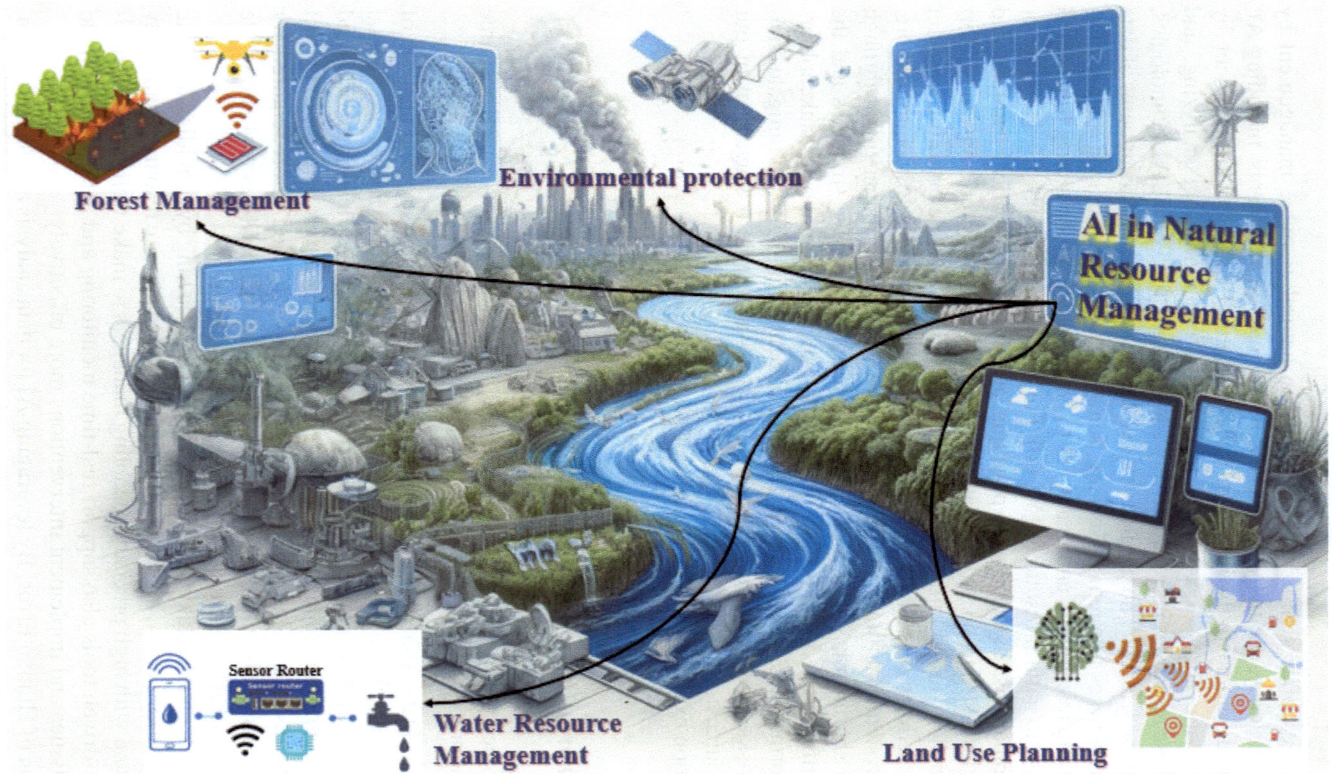

Fig. 4 Applications of artificial intelligence in natural resource management include monitoring forests, reducing deforestation, and supporting decision-makers in issuing early fire warnings. AI plays a crucial role in intelligent ecosystem restoration, adaptation to climate change, and minimizing ecosystem pollution while implementing effective conservation strategies. Additionally, AI is essential for the efficient management of urban water resources. Remote sensing and GIS technologies enhance urban land use and planning

manual processes. This leads to more reliable outcomes in resource management and conservation efforts.

3.2 Enhanced Data Utilization

AI enables the integration and comprehensive analysis of diverse data sources, leading to a better understanding of natural resource dynamics.

- Integration of Data Sources: AI technologies can combine data from various sources, such as remote sensing, ground-based sensors, and historical records [67, 68]. Techniques like data fusion and integration algorithms allow for the synthesis of disparate datasets into a unified view, providing a holistic understanding of resource conditions and trends.
- Comprehensive Analysis: Machine learning models and AI algorithms analyze complex datasets to uncover patterns and relationships that may not be evident through traditional methods. For example, AI can identify correlations between environmental factors and resource changes, leading to deeper insights into ecosystem dynamics and resource interactions [69].
- Real-Time Data Processing: AI enables the processing of real-time data streams from sensors and monitoring systems [70]. This capability allows for continuous tracking of resource conditions, immediate detection of anomalies, and prompt response to emerging issues. Real-time analysis enhances the ability to manage resources dynamically and adapt to changing conditions.

3.3 Cost Savings

AI-driven automation and predictive modeling result in significant cost savings for resource management and conservation.

- Reduction of Labor Costs: By automating data collection, analysis, and reporting, AI reduces the need for manual labor and associated costs [71, 72]. For example, automated monitoring systems and AI-powered analysis tools decrease the reliance on human resources for routine tasks, leading to operational cost savings.
- Optimized Resource Allocation: Predictive models and optimization algorithms help in allocating resources more effectively [73]. AI can identify areas where resources are most needed, optimize distribution strategies, and reduce waste. This optimization leads to more efficient use of resources and cost reductions in areas such as water distribution, energy management, and mineral extraction.
- Minimized Environmental Impact: AI-driven solutions can also contribute to cost savings by minimizing the environmental impact of resource management activities [74–76]. For instance, precision agriculture techniques powered by AI reduce the need for excessive use of fertilizers and pesticides, lowering costs and environmental degradation.

3.4 Increased Accessibility

AI tools democratize access to advanced analytical capabilities, making them available to a broader range of stakeholders.

- Support for Small-Scale Managers: AI technologies provide small-scale resource managers, such as local conservationists and community-based organizations, with access to sophisticated tools and data analytics [77]. This empowerment enables these stakeholders to make informed decisions and implement effective management practices without requiring extensive technical expertise.
- Enhanced Collaboration: AI facilitates collaboration among various stakeholders, including governments, NGOs, research institutions, and private enterprises [78]. By providing shared access to data and analytical tools, AI promotes collective efforts in resource management and conservation, leading to more coordinated and effective actions [79].
- Scalability of Solutions: AI solutions are scalable and can be adapted to different contexts and scales, from local to global levels. This scalability allows for the application of AI technologies in diverse settings, including urban areas, rural communities, and large-scale ecosystems, enhancing resource management across various contexts [80–82].

4 Challenges and Considerations

Despite the transformative potential of AI in natural resource management, several challenges and considerations must be addressed to ensure effective implementation and maximize benefits. This section explores these challenges in detail.

4.1 Data Quality and Availability

The effectiveness of AI in natural resource management heavily relies on the quality and availability of data.

- Data Accuracy and Completeness: AI models depend on accurate and comprehensive data to make reliable predictions and recommendations. Inaccurate, outdated, or incomplete data can lead to erroneous conclusions and ineffective management strategies. For example, erroneous sensor data or missing historical records can undermine the performance of predictive models and impact decision-making [83, 84].
- Data Integration: Integrating data from various sources, such as satellite imagery, ground-based sensors, and historical databases, presents a challenge [85]. Ensuring that data from diverse sources is compatible and accurately combined is

crucial for comprehensive analysis. Inconsistent data formats, spatial resolutions, and temporal scales can complicate data integration and analysis [86].
- Data Access and Sharing: Access to high-quality data can be restricted due to proprietary concerns, privacy regulations, or logistical constraints. Facilitating data sharing and ensuring that data is accessible to stakeholders while respecting privacy and intellectual property rights is essential for effective AI applications in resource management.

4.2 Ethical and Privacy Concerns

The application of AI in monitoring and managing natural resources raises several ethical and privacy issues.

- Surveillance and Privacy: AI technologies, such as remote sensing and automated monitoring systems, can lead to concerns about the surveillance of individuals or communities [87, 88]. The collection and analysis of data related to human activities, particularly in sensitive areas, must be conducted with caution to avoid infringing on privacy rights. Transparent policies and regulations are necessary to address these concerns and protect individuals' privacy.
- Bias and Fairness: AI systems can perpetuate or exacerbate existing biases if not carefully designed and monitored [89, 90]. Bias in training data or algorithmic decision-making can result in unfair or discriminatory outcomes, particularly in resource management contexts where certain groups may be disproportionately affected. Ensuring fairness and accountability in AI systems is critical to addressing these issues.
- Transparency and Accountability: The opacity of some AI models, particularly complex algorithms like deep learning networks, can hinder understanding and accountability [91]. Ensuring that AI systems are transparent and that their decision-making processes can be explained is important for building trust and accountability.

4.3 Integration with Existing Systems

Integrating AI technologies with existing resource management systems presents several challenges.

- Compatibility with Legacy Systems: Many resource management systems are based on legacy technologies that may not be compatible with modern AI tools. Integrating AI solutions with these systems requires careful planning and adaptation to ensure seamless interoperability. This may involve upgrading existing infrastructure or developing custom solutions to bridge technological gaps.

- Collaboration and Coordination: Effective integration of AI requires collaboration among AI experts, resource managers, and policymakers [92]. Interdisciplinary teams must work together to align AI solutions with resource management goals and practices. Coordinating efforts and establishing clear communication channels are essential for successful integration.
- Change Management: Implementing AI solutions often involves changes to established workflows and practices. Managing these changes requires addressing potential resistance from stakeholders, providing training, and ensuring that AI tools are user-friendly and aligned with users' needs.

4.4 Skill Requirements

The successful implementation and use of AI in natural resource management necessitates specialized skills and knowledge.

- Training and Capacity Building: Stakeholders, including resource managers and conservationists, need training to effectively use and interpret AI tools [39, 93]. Capacity building programs should focus on developing skills in data analysis, machine learning, and AI application in resource management. Ensuring that stakeholders have the necessary expertise to leverage AI tools is crucial for achieving desired outcomes.
- Interdisciplinary Knowledge: Effective use of AI in resource management requires knowledge from multiple disciplines, including computer science, environmental science, and domain-specific expertise [94]. Building interdisciplinary teams with diverse skill sets helps to address the complex challenges associated with AI applications.
- Ongoing Support and Education: As AI technologies evolve, ongoing support and education are necessary to keep stakeholders updated on new developments and best practices. Providing access to continuing education and professional development opportunities can help ensure that stakeholders remain proficient in using AI tools.

5 Future Directions

The future of AI in natural resource management is poised for significant advancements and innovations. As technology continues to evolve, several promising directions are emerging:

5.1 Advancements in AI Techniques

The evolution of AI techniques promises to further enhance the capabilities of AI in managing natural resources.

- **Deep Learning**: Advances in deep learning, particularly in neural network architectures, are expected to improve the accuracy and efficiency of AI models [95]. Techniques such as CNNs and RNNs are becoming more sophisticated, enabling more precise analysis of complex data sets [96]. These improvements will enhance the ability to monitor and predict changes in natural resources with greater accuracy.
- **Reinforcement Learning**: Reinforcement learning (RL) is gaining traction for its potential in optimizing resource management strategies [97]. RL algorithms can learn and adapt based on interactions with the environment, making them well-suited for dynamic and complex resource management scenarios. Future developments in RL could lead to more effective and adaptive management practices.
- **Explainable AI (XAI)**: As AI models become more complex, there is a growing need for explainable AI that provides transparent insights into decision-making processes. XAI aims to make AI systems more interpretable and accountable, which is crucial for gaining trust and ensuring that AI-driven decisions are understandable and justifiable [98].

5.2 Integration with IoT

The integration of AI with Internet of Things (IoT) technologies will revolutionize natural resource management by enabling real-time monitoring and control.

- **Real-Time Data Collection**: IoT devices, such as sensors and smart meters, will provide continuous, real-time data on various environmental parameters [99, 100]. AI can analyze this data to monitor resource conditions dynamically, detect anomalies, and respond to changes promptly. This integration will enhance the ability to manage resources in real-time, leading to more responsive and adaptive management strategies.
- **Smart Resource Management**: Combining AI with IoT will facilitate the development of smart resource management systems [75, 101]. For instance, smart irrigation systems can use AI to optimize water usage based on real-time soil moisture data and weather forecasts. Similarly, AI-powered energy management systems can optimize energy distribution and consumption based on real-time data from IoT sensors [102].
- **Predictive Maintenance**: IoT sensors can monitor the condition of infrastructure and equipment used in resource management [103]. AI can analyze this data to

predict maintenance needs and prevent failures before they occur. This predictive capability will reduce downtime and maintenance costs, improving overall resource management efficiency.

5.3 Collaboration and Partnerships

Collaboration among various stakeholders will be crucial for advancing AI solutions in sustainable resource management.

- **Governments and Policymakers**: Governments play a critical role in setting regulations and policies that support the deployment of AI in resource management [104]. Collaboration with AI experts can help in developing policies that facilitate innovation while addressing ethical and privacy concerns.
- **Academia and Research Institutions**: Academic and research institutions are essential for advancing AI research and developing new techniques [105]. Partnerships with these institutions can drive innovation and ensure that AI solutions are based on the latest scientific knowledge.
- **Industry and Technology Providers**: Collaboration with industry and technology providers will be vital for translating research into practical applications. Industry partnerships can help in scaling AI solutions and integrating them into existing systems and practices.
- **Non-Governmental Organizations (NGOs)**: NGOs or local communities often have valuable insights into local resource management challenges and can provide on-the-ground perspectives [106]. Collaborating with NGOs can ensure that AI solutions are tailored to address specific needs and contexts, promoting more effective and inclusive resource management.

6 Conclusion

This chapter has explored the multifaceted applications of Artificial Intelligence in Natural Resource Management, highlighting its transformative potential across diverse domains, including water resource management, forest management, mineral and mining management, and biodiversity conservation. The sources emphasize that AI, coupled with technologies like remote sensing, IoT, and big data analytics, empowers stakeholders with advanced tools for monitoring, analysis, and decision-making, leading to improved efficiency, accuracy, and sustainability in managing our planet's resources. The sources illustrate how AI algorithms, such as machine learning and deep learning models, excel in analysing complex datasets, extracting actionable insights, and making accurate predictions, enabling more informed and timely interventions. From predicting water demand and flood events to detecting deforestation and assessing forest health, AI offers innovative solutions for addressing critical challenges in NRM Moreover, AI-powered automation reduces reliance on manual labour, minimizes human error, optimizes resource allocation, and facilitates cost savings, further bolstering the case for its widespread adoption in NRM. However, the sources also acknowledge the challenges and considerations associated with AI implementation in NRM. Data quality and availability, ethical and privacy concerns, integration with existing systems, and skill requirements are crucial aspects that need careful attention. Addressing these challenges requires collaborative efforts among researchers, policymakers, industry experts, and local communities to ensure responsible and ethical AI development, promoting transparency, fairness, and accountability in its application. As AI technologies continue to evolve, their integration into NRM practices, guided by a commitment to sustainability and ethical considerations, holds immense promise for securing a future where our planet's resources are managed wisely and equitably.

References

1. Reichstein, M., Camps-Valls, G., Stevens, B., Jung, M., Denzler, J., Carvalhais, N., & Prabhat, F. (2019). Deep learning and process understanding for data-driven Earth system science. *Nature, 566*(7743), 195–204.
2. Hao, T., Elith, J., Guillera-Arroita, G., & Lahoz-Monfort, J. J. (2019). A review of evidence about use and performance of species distribution modelling ensembles like BIOMOD. *Diversity and Distributions, 25*(5), 839–852.
3. Kratzert, F., Klotz, D., Brenner, C., Schulz, K., & Herrnegger, M. (2018). Rainfall–runoff modelling using long short-term memory (LSTM) networks. *Hydrology and Earth System Sciences, 22*(11), 6005–6022.
4. Carranza, E. J. M., & Laborte, A. G. (2015). Random forest predictive modeling of mineral prospectivity with small number of prospects and data with missing values in Abra (Philippines). *Computers & Geosciences, 74*, 60–70.
5. Katkani, D., Babbar, A., Mishra, V. K., Trivedi, A., Tiwari, S., & Kumawat, R. K. (2022). A review on applications and utility of remote sensing and geographic information systems in agriculture and natural resource management. *International Journal of Environment and Climate Change, 12*(4), 1–18.
6. Fayaz, M., Nam, J., Dang, L. M., Song, H.-K., & Moon, H. (2024). Land-cover classification using deep learning with high-resolution remote-sensing imagery. *Applied Sciences, 14*(5), 1844.
7. Doull, K. E., Chalmers, C., Fergus, P., Longmore, S., Piel, A. K., & Wich, S. A. (2021). An evaluation of the factors affecting 'poacher'detection with drones and the efficacy of machine-learning for detection. *Sensors, 21*(12), 4074.
8. Teixeira, K., Miguel, G., Silva, H. S., & Madeiro, F. (2023). A survey on applications of unmanned aerial vehicles using machine learning. IEEE Access.
9. Munthali, M. G. (2020). *Analysis of land use and land cover dynamics and its implications on natural resources in Dedza District Malawi*. University of Pretoria (South Africa).
10. Haq, B., Jamshed, M. A., Ali, K., Kasi, B., Arshad, S., Kasi, M. K., ... Ur-Rehman, M. (2024). Tech-driven forest conservation: Combating deforestation with internet of things, artificial intelligence, and remote sensing. *IEEE Internet of Things Journal, 11*(14), 24551–24568.
11. Oliver, N. (2022). Artificial intelligence for social good-the way forward. In *Science, research and innovation performance of the EU 2022 Report* (pp. 604–707).
12. Tran Anh, D., Van, S. P., Dang, T. D., & Hoang, L. P. (2019). Downscaling rainfall using deep learning long short-term memory and feedforward neural network. *International Journal of Climatology, 39*(10), 4170–4188.

13. Kramar, V., & Alchakov, V. (2023). Time-series forecasting of seasonal data using machine learning methods. *Algorithms, 16*(5), 248.
14. Choi, K., Yi, J., Park, C., & Yoon, S. (2021). Deep learning for anomaly detection in time-series data: Review, analysis, and guidelines. *IEEE Access, 9*, 120043–120065.
15. Vélez, J., Castiblanco-Camacho, P. J., Tabak, M. A., Chalmers, C., Fergus, P., & Fieberg, J. (2022). *Choosing an appropriate platform and workflow for processing camera trap data using artificial intelligence.* arXiv preprint arXiv:2202.02283
16. Nguyen, H., Maclagan, S. J., Nguyen, T. D., Nguyen, T., Flemons, P., Andrews, K. … Phung, D. (2017). Animal recognition and identification with deep convolutional neural networks for automated wildlife monitoring. In *2017 IEEE international conference on data science and advanced Analytics (DSAA)*. IEEE.
17. Wong, W. Y., Al-Ani, A. K. I., Hasikin, K., Khairuddin, A. S. M., Razak, S. A., Hizaddin, H. F., … Azizan, M. M. (2021). Water, soil and air pollutants' interaction on mangrove ecosystem and corresponding artificial intelligence techniques used in decision support systems—a review. *IEEE Access, 9*, 105532–105563.
18. Popescu, S. M., Mansoor, S., Wani, O. A., Kumar, S. S., Sharma, V., Sharma, A., … Bolan, N. (2024). Artificial intelligence and IoT driven technologies for environmental pollution monitoring and management. *Frontiers in Environmental Science, 12*, 1336088.
19. Ye, Z., Yang, J., Zhong, N., Tu, X., Jia, J., & Wang, J. (2020). Tackling environmental challenges in pollution controls using artificial intelligence: A review. *Science of the Total Environment, 699*, 134279.
20. Pouyanfar, N., S.Z. Harofte, M. Soltani, S. Siavashy, E. Asadian, F. Ghorbani-Bidkorbeh, . . . C.M. Hussain, Artificial intelligence-based microfluidic platforms for the sensitive detection of environmental pollutants: Recent advances and prospects. Trends in Environmental Analytical Chemistry, 2022. 34: p. e00160.
21. Emvoliadis, A., Angelis, G.-F., Drosou, A., & Tzovaras, D. (2024). Improving air quality data analysis by injecting and detecting contextual anomalies. In *IGARSS 2024-2024 IEEE International Geoscience and Remote Sensing Symposium*. IEEE.
22. Abid, S. K., Sulaiman, N., Chan, S. W., Nazir, U., Abid, M., Han, H., … Vega-Muñoz, A. (2021). Toward an integrated disaster management approach: How artificial intelligence can boost disaster management. *Sustainability, 13*(22), 12560.
23. Zhu, M., Wang, J., Yang, X., Zhang, Y., Zhang, L., Ren, H., … Ye, L. (2022). A review of the application of machine learning in water quality evaluation. *Eco-Environment & Health, 1*(2), 107–116.
24. Shams, M. Y., Elshewey, A. M., El-Kenawy, E.-S.M., Ibrahim, A., Talaat, F. M., & Tarek, Z. (2024). Water quality prediction using machine learning models based on grid search method. *Multimedia Tools and Applications, 83*(12), 35307–35334.
25. Hmoud Al-Adhaileh, M., & Waselallah Alsaade, F. (2021). Modelling and prediction of water quality by using artificial intelligence. *Sustainability, 13*(8), 4259.
26. Rankinen, K., Bernal, J. E. C., Holmberg, M., Vuorio, K., & Granlund, K. (2019). Identifying multiple stressors that influence eutrophication in a Finnish agricultural river. *Science of the Total Environment, 658*, 1278–1292.
27. Ma, J., Ding, Y., Cheng, J. C., Jiang, F., & Xu, Z. (2020). Soft detection of 5-day BOD with sparse matrix in city harbor water using deep learning techniques. *Water research, 170*, 115350.
28. Rajesh, M., & Rehana, S. (2021). Prediction of river water temperature using machine learning algorithms: A tropical river system of India. *Journal of Hydroinformatics, 23*(3), 605–626.
29. Chopade, S., Gupta, H. P., Mishra, R., Oswal, A., Kumari, P., & Dutta, T. (2021). A sensors-based river water quality assessment system using deep neural network. *IEEE Internet of Things Journal, 9*(16), 14375–14384.
30. Bi, J., Lin, Y., Dong, Q., Yuan, H., & Zhou, M. (2021). Large-scale water quality prediction with integrated deep neural network. *Information Sciences, 571*, 191–205.
31. Kalantar, B., Ueda, N., Idrees, M. O., Janizadeh, S., Ahmadi, K., & Shabani, F. (2020). Forest fire susceptibility prediction based on machine learning models with resampling algorithms on remote sensing data. *Remote Sensing, 12*(22), 3682.

32. Onifade, M., Adebisi, J. A., Shivute, A. P., & Genc, B. (2023). Challenges and applications of digital technology in the mineral industry. *Resources Policy, 85*, 103978.
33. Ali, D., & Frimpong, S. (2020). Artificial intelligence, machine learning and process automation: Existing knowledge frontier and way forward for mining sector. *Artificial Intelligence Review, 53*(8), 6025–6042.
34. Molaei, F., Rahimi, E., Siavoshi, H., Afrouz, S. G., & Tenorio, V. (2020). A comprehensive review on internet of things (IoT) and its implications in the mining industry. *American Journal of Engineering and Applied Sciences, 13*(3), 499–515.
35. Hussain, R. F., Mokhtari, A., Ghalambor, A., & Salehi, M. A. (2022). *IoT for smart operations in the oil and gas industry: From upstream to downstream.* Gulf Professional Publishing.
36. Ramezani, M., & Tafazoli, S. (2020). Using artificial intelligence in mining excavators: Automating routine operational decisions. *IEEE Industrial Electronics Magazine, 15*(1), 6–11.
37. Rourke, M. L., Fowler, A. M., Hughes, J. M., Broadhurst, M. K., DiBattista, J. D., Fielder, S., ... Furlan, E. M. (2022). Environmental DNA (eDNA) as a tool for assessing fish biomass: A review of approaches and future considerations for resource surveys. *Environmental DNA, 4*(1), 9–33.
38. Ditria, E. M., Buelow, C. A., Gonzalez-Rivero, M., & Connolly, R. M. (2022). Artificial intelligence and automated monitoring for assisting conservation of marine ecosystems: A perspective. *Frontiers in Marine Science, 9*, 918104.
39. Shivaprakash, K. N., Swami, N., Mysorekar, S., Arora, R., Gangadharan, A., Vohra, K., ... Kiesecker, J. M. (2022). Potential for artificial intelligence (AI) and machine learning (ML) applications in biodiversity conservation, managing forests, and related services in India. *Sustainability, 14*(12), 7154.
40. Carreiras, J., Pereira, J., & Shimabukuro, Y. E. (2006). Land-cover mapping in the Brazilian Amazon using SPOT-4 vegetation data and machine learning classification methods. *Photogrammetric Engineering & Remote Sensing, 72*(8), 897–910.
41. Giannetti, F., Barbati, A., Mancini, L. D., Travaglini, D., Bastrup-Birk, A., Canullo, R., ... Chirici, G. (2018). European forest types: Toward an automated classification. *Annals of Forest Science, 75*, 1–14.
42. Lin, P., Lu, Q., Li, D., Chen, Y., Zou, Z., & Jiang, S. (2020). Artificial intelligence classification of wetland vegetation morphology based on deep convolutional neural network. *Natural Resource Modeling, 33*(1), e12248.
43. Watanabe, S., Sumi, K., & Ise, T. (2018). Automatic vegetation identification in Google Earth images using a convolutional neural network: A case study for Japanese bamboo forests. *bioRxiv 351643.*
44. Bastin, J.-F., Finegold, Y., Garcia, C., Mollicone, D., Rezende, M., Routh, D., ... Crowther, T. W. (2019). The global tree restoration potential. *Science, 365*(6448), 76–79.
45. Arekhi, S., & Jafarzadeh, A. A. (2014). Forecasting areas vulnerable to forest conversion using artificial neural network and GIS (case study: Northern Ilam forests, Ilam province, Iran). *Arabian Journal of Geosciences, 7*(3), 1073–1085.
46. Mayfield, H. J., Smith, C., Gallagher, M., & Hockings, M. (2020). Considerations for selecting a machine learning technique for predicting deforestation. *Environmental Modelling & Software, 131*, 104741.
47. da Rocha, S. J. S. S., Torres, C. M. M. E., Jacovine, L. A. G., Leite, H. G., Gelcer, E. M., Neves, K. M., ... Reis, L. P. (2018). Artificial neural networks: Modeling tree survival and mortality in the Atlantic Forest biome in Brazil. *Science of the Total Environment, 645*, 655–661.
48. Satir, O., Berberoglu, S., & Donmez, C. (2016). Mapping regional forest fire probability using artificial neural network model in a Mediterranean forest ecosystem. *Geomatics, Natural Hazards and Risk, 7*(5), 1645–1658.
49. Rammer, W., & Seidl, R. (2019). Harnessing deep learning in ecology: An example predicting bark beetle outbreaks. *Frontiers in Plant Science, 10*, 451705.
50. Crunchant, A.-S., Nakka, C. D., Isaacs, J. T., & Piel, A. K. (2021). Acoustic sensors. In *Conservation Technology* (p. 53).

51. Kuruppu, S. (2023). AI system to protect endangered animgal population and prevent poaching threats using weapon detection. *International Journal of Innovative Science and Research Technology, 8*(9), 1270–1275.
52. Raihan, A. (2023). Artificial intelligence and machine learning applications in forest management and biodiversity conservation. *Natural Resources Conservation and Research, 6*(2), 3825.
53. Radha, M., Kannan, A. Priyadharshini, K., Krishnarani, M., Sankar, G. R., & Narayanan, K. L. (2023). Sustainable forest monitoring: A comprehensive wireless sensor solution against illegal logging. In *2023 International Conference on Self Sustainable Artificial Intelligence Systems (ICSSAS)*. IEEE.
54. Sultana, M., Bhattacharya, S., Adhikari, N., Sengupta, D., De, D. (2024). Artificial intelligence for sustainable. In *Artificial intelligence and edge computing for sustainable ocean health* (p. 3).
55. Schrodt, F., De La Barreda Bautista, B., Williams, C., Boyd, D. S., Schaepman-Strub, G., Santos, M. J. (2020). Integrating biodiversity, remote sensing, and auxiliary information for the study of ecosystem functioning and conservation at large spatial scales. In *Remote sensing of plant biodiversity* (pp. 449–484).
56. Buchelt, A., Adrowitzer, A., Kieseberg, P., Gollob, C., Nothdurft, A., Eresheim, S., … Holzinger, A. (2024). Exploring artificial intelligence for applications of drones in forest ecology and management. *Forest Ecology and Management, 551*, 121530.
57. Carayannis, E. G., Dezi, L., Gregori, G., & Calo, E. (2021). Smart environments and techno-centric and human-centric innovations for Industry and Society 5.0: A quintuple helix innovation system view towards smart, sustainable, and inclusive solutions. *Journal of the Knowledge Economy*, 1–30.
58. Melnychenko, S., Cholyshkina, O., Rtyshchev, S., Mishchenko, A., Karpinskyi, R. (2024). Intelligent systems in the management of regional resources: Benefits for sustainable development. *Studia Ecologiae et Bioethicae, 22*(3).
59. Khoulji, S., Chouya, S., Bourekkadi, S., Slimani, K., & Kerkeb, M. (2017). Intelligent system for the management of resources dedicated to humanity. *Transactions on Machine Learning and Artificial Intelligence, 5*(4).
60. Kovalenko, Y., Zakharova, O., Yakusheva, O., Yakushev, O., Hulak, D., & Kozlovska, S. (2021). Priorities for the development of regional information management based on system digitalization. In *SHS Web of Conferences*. EDP Sciences.
61. Samara, E., Andronikidis, A., Komninos, N., Bakouros, Y., & Katsoras, E. (2023). The role of digital technologies for regional development: A system dynamics analysis. *Journal of the Knowledge Economy, 14*(3), 2215–2237.
62. Kamyab, H., Khademi, T., Chelliapan, S., SaberiKamarposhti, M., Rezania, S., Yusuf, M., … Ahn, Y. (2023). The latest innovative avenues for the utilization of artificial Intelligence and big data analytics in water resource management. *Results in Engineering, 20*, 101566.
63. Kilintzis, P., Samara, E., Carayannis, E. G., & Bakouros, Y. (2020). Business model innovation in Greece: Its effect on organizational sustainability. *Journal of the Knowledge Economy, 11*, 949–967.
64. Kanga, S. (2023). Advancements in remote sensing tools for forestry analysis. *Sustainable Forestry, 6*(1).
65. Acharjee, P. (2024). AI in environmental applications. In *Artificial intelligence for multimedia information processing* (pp. 195–218). CRC Press.
66. Oliff, H., Liu, Y., Kumar, M., & Williams, M. (2020). Improving human–robot interaction utilizing learning and intelligence: A human factors-based approach. *IEEE Transactions on Automation Science and Engineering, 17*(3), 1597–1610.
67. Yang, L., Driscol, J., Sarigai, S., Wu, Q., Chen, H., & Lippitt, C. D. (2022). Google Earth Engine and artificial intelligence (AI): A comprehensive review. *Remote Sensing, 14*(14), 3253.
68. Janga, B., Asamani, G. P., Sun, Z., & Cristea, N. (2023). A review of practical ai for remote sensing in earth sciences. *Remote Sensing, 15*(16), 4112.

69. Liu, Y., Huang, X., & Liu, Y. (2024). Detection of long-term land use and ecosystem services dynamics in the Loess Hilly-Gully region based on artificial intelligence and multiple models. *Journal of Cleaner Production, 447*, 141560.
70. Habeeb, R. A. A., Nasaruddin, F., Gani, A., Hashem, I. A. T., Ahmed, E., & Imran, M. (2019). Real-time big data processing for anomaly detection: A survey. *International Journal of Information Management, 45*, 289–307.
71. Samokhvalov, I. (2024). Transforming management reporting with intelligent process automation (IPA): Enhancing business analytics, forecasting, and decision-making in organizations.
72. Jędrzejka, D. (2019). Robotic process automation and its impact on accounting. *Zeszyty Teoretyczne Rachunkowości, 105*, 137–166.
73. Nicklow, J., Reed, P., Savic, D., Dessalegne, T., Harrell, L., Chan-Hilton, A., … Singh, A. (2010). State of the art for genetic algorithms and beyond in water resources planning and management. *Journal of Water Resources Planning and Management, 136*(4), 412–432.
74. Reza, M., & Hassan, M. (2023). AI-Driven solutions for enhanced waste management and recycling in urban areas. *International Journal of Sustainable Infrastructure for Cities and Societies, 8*(2), 1–13.
75. Kothamali, P. R., Mandaloju, N., & Dandyala, S. S. M. (2022). Optimizing resource management in smart cities with AI. *Unique Endeavor in Business & Social Sciences, 1*(1), 174–191.
76. Ming, L. W., Anderson, J., Hidayat, F., Yulian, F. D., & Septiani, N. (2024). Ai as a driver of efficiency in waste management and resource recovery. *International Transactions on Artificial Intelligence, 2*(2), 128–134.
77. Karl, S., Er, Z. Geographical perspectives on 1 globalization and environmental issues: The inner-connections of conservation.
78. Charalabidis, Y. (2022). Artificial intelligence (AI). *The Palgrave encyclopedia of interest groups, lobbying and public affairs* (pp. 43–50). Springer.
79. Allam, Z. (2020). *The rise of autonomous smart cities: Technology, economic performance and climate resilience*. Springer Nature.
80. Yigitcanlar, T., David, A., Li, W., Fookes, C., Bibri, S. E., & Ye, X. (2024). Unlocking artificial intelligence adoption in local governments: Best practice lessons from real-world implementations. *Smart Cities, 7*(4), 1576–1625.
81. Kaur, G., Tomar, P., & Tanque, M. (2020). *Artificial intelligence to solve pervasive internet of things issues*. Academic Press.
82. Hamerlinck, J. D. (2020). Applying planning support science in rural environments. In *Handbook of planning support science* (pp. 524–538). Edward Elgar Publishing.
83. Marr, B. (2015). *Big data: Using SMART big data, analytics and metrics to make better decisions and improve performance*. Wiley.
84. Guimerà Cuevas, F. (2024). *Artificial Intelligence for automated decision-making in error-pattern recognition*. LMU.
85. Ang, L.-M., Seng, K. P., Zungeru, A. M., & Ijemaru, G. K. (2017). Big sensor data systems for smart cities. *IEEE Internet of Things Journal, 4*(5), 1259–1271.
86. Verburg, P. H., Neumann, K., & Nol, L. (2011). Challenges in using land use and land cover data for global change studies. *Global Change Biology, 17*(2), 974–989.
87. Wiesenthal, M. (2022). *The ethical implications of AI-based mass surveillance tools*.
88. Räty, T. D. (2010). Survey on contemporary remote surveillance systems for public safety. *IEEE Transactions on Systems, Man, and Cybernetics, Part C (Applications and Reviews), 40*(5), 493–515.
89. Mensah, G. B. (2023). *Artificial intelligence and ethics: a comprehensive review of bias mitigation, transparency, and accountability in AI Systems*. Preprint, November 10, 2023.
90. Min, A. (2023). Artificial intelligence and bias: Challenges, implications, and remedies. *Journal of Social Research, 2*(11).
91. Beaudouin, V., Bloch, I., Bounie, D., Clémençon, S., d'Alché-Buc, F., Eagan, J., … Parekh, J. (2020). *Flexible and context-specific AI explainability: A multidisciplinary approach*. arXiv preprint arXiv:2003.07703.

92. Chen, Y.-C., Liu, H., & Wang, Y.-F. (2024). Governance design of collaborative intelligence for public policy and services. In *Proceedings of the 25th Annual International Conference on Digital Government Research.*
93. van der Merwe, J., Wahid, S. M., Cesna, G. P., & Prabowo, D. A. (2024). Improving natural resource management through AI: Quantitative analysis using smartpls. *International Transactions on Artificial Intelligence, 2*(2), 135–142.
94. Konya, A., & Nematzadeh, P. (2024). Recent applications of AI to environmental disciplines: A review. *Science of The Total Environment, 906*, 167705.
95. Ahmed, S. F., Alam, M. S. B., Hassan, M., Rozbu, M. R., Ishtiak, T., Rafa, N., … Gandomi, A. H. (2023). Deep learning modelling techniques: Current progress, applications, advantages, and challenges. *Artificial Intelligence Review, 56*(11), 13521–13617.
96. Minoli, D., Occhiogrosso, B. (2023). *AI Applications to communications and information technologies: The role of ultra deep neural networks.* Wiley.
97. Cheng, N., Wang, X., Li, Z., Yin, Z., Luan, T., & Shen, X. S. (2024). *Toward enhanced reinforcement learning-based resource management via digital twin: Opportunities, applications, and challenges.* IEEE Network.
98. Somani, A., Horsch, A., & Prasad, D. K. (2023). *Interpretability in deep learning* (Vol. 2). Springer.
99. Lezzar, F., Benmerzoug, D., & Kitouni, I. (2020). *IoT for monitoring and control of water quality parameters.*
100. Vermesan, O., & Friess, P. (2013). *Internet of things: converging technologies for smart environments and integrated ecosystems.* River Publishers.
101. Walia, G. K., Kumar, M., & Gill, S. S. (2023). AI-empowered fog/edge resource management for IoT applications: A comprehensive review, research challenges and future perspectives. IEEE Communications Surveys & Tutorials.
102. Raj, N., Sinha, A., Bharddwaj, S., & Yadav, A. L. (2024). AI-powered energy consumption optimization for smart homes using IoT. In *2024 International Conference on Computational Intelligence and Computing Applications (ICCICA).* IEEE.
103. Maraveas, C., Piromalis, D., Arvanitis, K. G., Bartzanas, T., & Loukatos, D. (2022). Applications of IoT for optimized greenhouse environment and resources management. *Computers and Electronics in Agriculture, 198*, 106993.
104. Renda, A. (2024). Governance, government records, and the policymaking process aided by AI. In *Handbook of artificial intelligence at work* (pp. 291–313). Edward Elgar Publishing.
105. Amini, L., Chen, C.-H., Cox, D., Oliva, A., & Torralba, A. (2020). Experiences and insights for collaborative industry-academic research in artificial intelligence. *AI Magazine, 41*(1), 70–81.
106. Roka, K. (2020). Community-based natural resources management. *Life on land* (pp. 161–174). Springer.

Chapter 6
AI for Natural Disaster Prediction and Management

Balendra V. S. Chauhan, Ajitanshu Vedrtnam, Kevin P. Wyche, and Sneha Verma

Abstract This chapter explores the transformative potential of Artificial Intelligence (AI) in predicting and managing natural disasters, focusing on earthquakes, floods, hurricanes, typhoons, and volcanic eruptions. The chapter reviews various AI techniques employed in disaster management, including machine learning models for forecasting, pattern recognition and anomaly detection, integration of AI with existing monitoring systems, and predictive models for aftershocks. It emphasizes the role of AI in enhancing traditional disaster management practices by improving the accuracy, timeliness, and efficiency of prediction, monitoring, and response efforts. The chapter discusses the integration of AI with remote sensing technologies, highlighting its effectiveness in analyzing satellite imagery and aerial photography for rapid damage assessment, flood mapping, and storm tracking. Real-time monitoring and early warning systems powered by AI are examined, emphasizing their potential to mitigate the impact of flood events and volcanic eruptions by providing timely alerts based on integrated data from diverse sources. Challenges and future directions for AI in natural disaster management are also addressed, including data quality and availability, model interpretability, and the need for localized calibration of AI models to account for regional variations. Future research directions are highlighted,

A. Vedrtnam
Mineral and Energy Economy Institute, Polish Academy of Sciences, Kraków, Poland
e-mail: ajitanshu.m@invertis.org

B. V. S. Chauhan (✉) · K. P. Wyche
Centre for Earth Observation Science, University of Brighton, Brighton, UK
e-mail: balendrachauhan8801@gmail.com

K. P. Wyche
e-mail: kevinpwyche@googlemail.com

B. V. S. Chauhan · S. Verma
School of Engineering, Newcastle University, Newcastle Upon Tyne, United Kingdom
e-mail: Sneha.verma@newcastle.ac.uk

A. Vedrtnam
Institute of Construction and Architecture, Slovak Academy of Science, 84503 Bratislava, Slovakia

Department of Mechanical Engineering, Invertis University, Bareilly 243001, UP, India

A. Vedrtnam et al., *Prospects of Artificial Intelligence in the Environment*,
https://doi.org/10.1007/978-981-96-6863-2_6

focusing on the integration of AI with emerging technologies like the Internet of Things (IoT), leveraging advanced data analytics, fostering global collaboration and knowledge sharing, and establishing public–private partnerships to drive innovation in AI solutions for disaster management.

Keywords Artificial Intelligence · Natural disaster management · Earthquake prediction · Flood forecasting · Hurricane monitoring · Typhoon tracking · Volcanic eruption prediction · AI in remote sensing · Early warning systems · Disaster risk assessment

1 Introduction

Artificial intelligence (AI) has emerged as a powerful tool in various fields, including the critical area of natural disaster prediction and management. AI refers to the capacity of digital systems to perform tasks that typically require human intelligence, encompassing a wide range of technologies including machine learning, deep learning, computer vision, and natural language processing. As the frequency and severity of natural disasters increase worldwide, often attributed to climate change, the application of AI in disaster management has become increasingly important. The use of AI in disaster management spans the entire cycle of mitigation, preparedness, response, and recovery [1]. AI technologies have shown significant potential in various applications, such as predicting and detecting geohazards, developing early warning systems (EWSs), assessing vulnerability and risk, and creating spatial models for mitigation strategies [2]. For instance, AI-based flood forecasting systems have proven essential for managing contingencies and emergencies effectively, mitigating risks, and reducing loss of life and property [3]. AI techniques are also being utilized for drought prediction across different timeframes, enabling better preparedness and mitigation strategies [4]. In the context of biological disasters, AI has played a crucial role in predicting disease outbreaks, developing diagnostic tools, and combating misinformation, as evidenced during the COVID-19 pandemic [5]. While AI offers transformative potential in disaster management, it also faces several challenges that must be addressed to ensure its responsible and effective use. These include issues of bias, discrimination, lack of transparency, data privacy and security, reliability, and ethical considerations. The concept of "trustworthy AI" has gained prominence, emphasizing the need for AI systems that are not only effective but also aligned with ethical principles and societal values [6]. To establish trustworthy AI in disaster management, several key factors must be considered [7]. These include robustness and reliability, privacy and security measures, fairness and bias mitigation, transparency and interoperability, and human oversight and accountability [8]. Organizations such as UNESCO and the European Union are leading efforts to establish ethical guidelines and legislative frameworks for AI deployment [9]. The field of Explainable Artificial Intelligence (XAI) has seen significant growth, particularly in its application to natural disasters. XAI techniques, such as SHAP (SHapley

Additive exPlanations), Grad-CAM (Gradient-weighted Class Activation Mapping), and LIME (Local Interpretable Model-agnostic Explanations), are being employed to enhance the transparency and interpretability of AI models used in disaster prediction and management [10–12]. In the context of natural disasters, data fusion techniques play a vital role in integrating and analysing diverse datasets from various sources, such as satellite imagery, sensor networks, social media, and historical records [13]. By combining multiple data sources and leveraging AI algorithms, data fusion can enhance situational awareness, enable early warning systems, and support effective response strategies. As we continue to develop and refine AI applications for disaster management, it is crucial to adopt a multidisciplinary approach that combines technological innovation with domain expertise in disaster response, environmental science, and social sciences. By addressing the challenges and embracing the potential of AI, we can work towards more effective, efficient, and equitable disaster management strategies that save lives and protect vulnerable communities in the face of increasing natural hazards [14].

2 AI Techniques for Earthquake Prediction and Early Warning

Earthquakes represent a significant natural disaster challenge due to their sudden onset and devastating impact. Traditional seismic monitoring relies heavily on physical seismometers and networked sensor systems to detect and record earthquake activity. However, these methods alone are insufficient for predicting earthquakes with high accuracy and timeliness. AI techniques have emerged as powerful tools to enhance earthquake prediction and early warning systems, Fig. 1 shows the areas of applications of AI in earthquake mitigation.

2.1 Machine Learning Models for Earthquake Prediction

The application of machine learning (ML) techniques to earthquake prediction has gained significant traction in recent years, offering promising results in improving the accuracy and reliability of seismic forecasts. Various ML models have been employed to tackle this complex challenge, each with its own strengths and limitations. Random Forest (RF) models have shown particular promise in earthquake prediction tasks. Yavas et al. achieved a remarkable 97.97% accuracy in predicting the maximum earthquake category within a 30-day period for Los Angeles, California, using a Random Forest model [15]. Rouet-Leduc et al. demonstrated the potential of machine learning to predict fault failure times in laboratory experiments, achieving high accuracy using a Random Forest model [16]. This study highlighted the potential of ML techniques to significantly enhance earthquake prediction capabilities. This

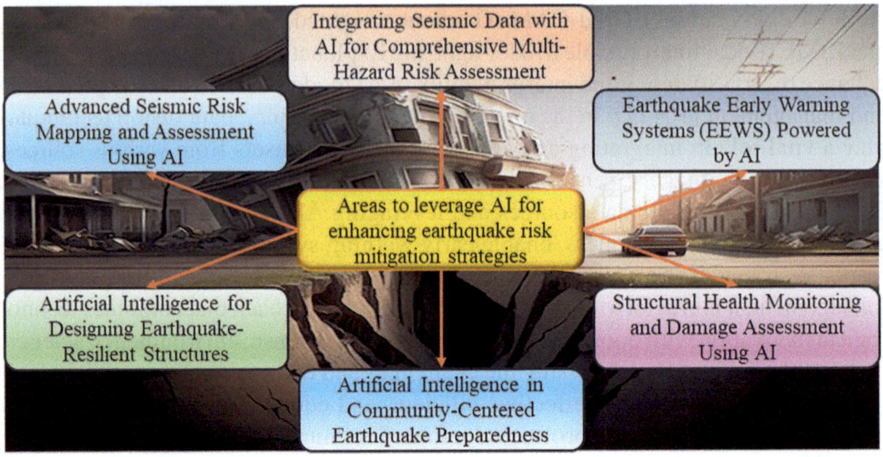

Fig. 1 Aspects of application of AI in earthquake risk mitigation

study demonstrated the potential of ML techniques to significantly enhance earthquake prediction capabilities, surpassing previous accuracy levels. Support Vector Machines (SVM) have also been applied successfully to earthquake prediction [17]. Asim et al. compared several ML techniques, including SVM, for predicting earthquake magnitudes in the Hindukush region, with SVM showing competitive performance [18]. Neural Networks (NN) have been widely explored for their ability to capture complex patterns in seismic data. Panakkat and Adeli used neural networks to predict the magnitude of the largest earthquake in the following month, demonstrating the potential of NNs in short-term earthquake forecasting [19].

More advanced deep learning techniques have also been applied to earthquake prediction. DeVries et al. developed a convolutional neural network approach for earthquake detection and location, showing improved performance over traditional methods [20]. Despite these promising results, challenges remain in the application of ML to earthquake prediction. Issues such as data quality, model interpretability, and the handling of imbalanced datasets continue to be areas of active research. Kong et al. provide a comprehensive review of these challenges and opportunities in applying machine learning to seismology [21]. As the field progresses, there is a growing emphasis on developing explainable AI (XAI) models to enhance trust and understanding in earthquake prediction systems. Techniques such as SHAP (SHapley Additive exPlanations) are being explored to provide insights into the decision-making processes of ML models in seismic contexts. In conclusion, while ML models have shown significant promise in improving earthquake prediction accuracy, continued research is needed to address remaining challenges and to develop robust, interpretable, and generalizable models for operational use in seismic risk management.

2.2 Pattern Recognition and Anomaly Detection

Pattern recognition and anomaly detection techniques have emerged as powerful tools in the field of earthquake prediction and early warning systems. These AI-based approaches aim to identify subtle patterns and anomalies in seismic data that may precede significant earthquake events. One of the most promising applications of pattern recognition in earthquake prediction is the analysis of seismic waveforms [22, 23]. Recent studies have demonstrated the effectiveness of convolutional neural networks (CNNs) in detecting and classifying seismic events [24]. These approaches have achieved high accuracy in distinguishing between earthquakes and non-earthquake events, showcasing the potential of deep learning techniques in automated seismic monitoring. According to a study by Jiao and Shan, anomaly detection methods have shown promise in identifying precursory signals that may indicate impending seismic activity [25]. Research applying machine learning techniques to laboratory earthquake data has successfully predicted the timing of lab-simulated earthquakes. A study by Rouet et al. demonstrated that machine learning algorithms could identify subtle statistical patterns in acoustic data that humans might overlook, opening new avenues for real-world earthquake prediction [16]. Recent advancements in AI have enabled the integration of multiple data sources for more comprehensive earthquake prediction models. Novel deep learning frameworks that combine seismic, geodetic, and geomagnetic data for earthquake prediction have been proposed. These approaches have demonstrated improved accuracy in forecasting seismic events compared to traditional methods, highlighting the potential of multi-modal data fusion in earthquake prediction.

Despite these promising results, challenges remain in applying pattern recognition and anomaly detection to real-world earthquake prediction [26]. Comprehensive reviews of machine learning applications in seismology highlight both the potential and limitations of these techniques. They emphasize the need for careful validation and interpretation of AI models, particularly when dealing with the complex and often unpredictable nature of seismic events. As the field progresses, there is a growing emphasis on developing explainable AI models to enhance trust and understanding in earthquake prediction systems [27]. The importance of explainable AI in critical applications like disaster management has been discussed in recent literature, proposing various techniques to make AI models more interpretable and trustworthy [10, 28–30]. Hence, pattern recognition and anomaly detection techniques show great promise in advancing earthquake prediction and early warning capabilities. However, continued research is needed to address challenges related to data quality, model interpretability, and the integration of AI systems with existing seismological frameworks.

2.3 Integration of AI with Seismic Monitoring Systems

The integration of artificial intelligence (AI) with seismic monitoring systems represents a significant advancement in earthquake prediction and early warning capabilities. This synergy enhances the accuracy, speed, and reliability of seismic data analysis, potentially leading to more timely and precise earthquake forecasts. Traditional seismic monitoring systems rely on networks of seismometers and accelerometers to detect ground motion. However, the vast amount of data generated by these networks can be challenging to process and interpret in real-time using conventional methods. AI techniques, particularly machine learning algorithms, have shown great promise in addressing these challenges. One notable application is the use of CNNs for rapid earthquake detection and location. Ross et al. demonstrated that a CNN-based model could accurately identify and locate earthquakes in Southern California, achieving performance comparable to that of human experts but at a much faster rate [31]. This approach significantly reduces the time required for initial earthquake detection, which is crucial for early warning systems. Deep learning models have also been applied to improve the accuracy of magnitude estimation. Mousavi and Beroza developed a deep neural network that can estimate earthquake magnitudes using only a few seconds of seismic waveform data [32]. Their model outperformed traditional methods, especially for smaller earthquakes, which are often challenging to characterize accurately. The integration of AI with seismic monitoring systems extends beyond single-station analysis. Perol et al. introduced ConvNetQuake, a convolutional neural network that can detect and locate earthquakes using a single ground-motion seismogram [33]. This approach has the potential to expand earthquake monitoring capabilities to areas with sparse seismic networks. AI techniques are also being used to enhance the discrimination between earthquakes and other seismic events, such as explosions or mining-induced tremors. Linville et al. employed machine learning algorithms to differentiate between natural earthquakes and human-induced seismic events in Oklahoma, achieving high accuracy in event classification [34]. Furthermore, the integration of AI with seismic monitoring systems has opened up new possibilities for earthquake forecasting. Rouet-Leduc et al. used machine learning to analyse continuous seismic data from a laboratory earthquake machine, successfully predicting the timing of laboratory-simulated earthquakes [16]. While the application of this approach to real-world scenarios remains challenging, it demonstrates the potential of AI in identifying precursory signals that may not be apparent through traditional analysis methods. As seismic monitoring networks continue to expand and generate increasingly large datasets, the role of AI in processing and interpreting this data is likely to grow. However, challenges remain, including the need for large, high-quality training datasets and the interpretability of complex AI models. Future research should focus on addressing these challenges and on developing AI systems that can integrate multiple data sources, including geodetic and geophysical data, to provide more comprehensive earthquake forecasts and early warnings.

2.4 Predictive Models for Aftershocks

Predicting aftershocks is a critical aspect of earthquake management. AI models can analyse data from initial earthquakes to forecast aftershock probabilities, improving the accuracy of aftershock forecasts and enhancing emergency response efforts. These models are crucial for developing strategies to reduce secondary damage and optimize resource allocation in the aftermath of a major seismic event. Recent advancements in machine learning have significantly improved our ability to predict aftershock patterns. DeVries et al. developed a deep learning model that outperformed traditional methods in forecasting aftershock locations [20]. Their neural network, trained on a database of over 131,000 mainshock-aftershock pairs, demonstrated a remarkable ability to identify complex patterns in static stress changes that are indicative of likely aftershock locations. Building on this work, Mignan and Broccardo proposed a hybrid model that combines machine learning with physical constraints [35]. Their approach, which integrates statistical seismology with deep learning, showed improved performance in aftershock forecasting while maintaining physical interpretability. This hybrid methodology addresses one of the key challenges in AI-based earthquake prediction: balancing the power of data-driven approaches with established geophysical principles. The application of AI to aftershock prediction extends beyond location forecasting. Liu et al. developed a machine learning model to predict the magnitude and timing of aftershocks [36]. Their approach, which utilizes a combination of seismic, geodetic, and geological data, demonstrated significant improvements in short-term aftershock forecasting accuracy compared to traditional statistical methods.

Real-time data processing and integration are crucial for effective aftershock prediction. Shen et al. proposed a deep learning framework that can rapidly process and analyse seismic waveforms to predict aftershock sequences in near real-time [37]. This approach has the potential to provide critical information to emergency responders and decision-makers in the immediate aftermath of a major earthquake. Despite these advancements, challenges remain in the field of AI-driven aftershock prediction. Zhang et al. highlighted the importance of model interpretability and uncertainty quantification in aftershock forecasting [38]. They proposed a novel deep learning Earthquake early warning approach that not only predicts aftershock patterns but also provides uncertainty estimates, which are crucial for risk assessment and decision-making in emergency situations. As AI techniques continue to evolve, there is growing interest in integrating these predictive models into operational earthquake early warning systems. Mousavi et al. demonstrated the potential of deep learning in real-time earthquake detection and characterization, which could be extended to aftershock prediction and integrated into existing early warning frameworks [39]. In conclusion, AI-driven predictive models for aftershocks represent a significant advancement in earthquake science and disaster management. These models offer the potential to improve the accuracy and timeliness of aftershock forecasts, ultimately contributing to more effective risk mitigation strategies and emergency response planning. However, continued research is needed to address challenges related to model interpretability, uncertainty quantification, and real-time implementation in operational settings.

2.5 Challenges and Future Directions

Despite the significant advancements in AI for earthquake prediction and early warning, several challenges persist that require ongoing research and development. These challenges span from data quality and quantity issues to the practical implementation of AI systems in operational environments. One of the primary challenges is the quality and quantity of seismic data required to train AI models effectively. Insufficient or biased data can lead to inaccuracies and unreliable predictions. Jia et al. highlight the importance of high-quality, diverse datasets in developing robust machine learning models for seismology [40]. They emphasize that the performance of AI models is heavily dependent on the comprehensiveness and accuracy of the training data. Another significant challenge is the integration of AI into operational earthquake early warning systems. This integration requires careful validation and testing to ensure reliability in real-world scenarios. Pwavodi et al. discuss the complexities of implementing deep learning models in operational settings, emphasizing the need for rigorous testing and continuous model updating to maintain performance over time [41]. The interpretability of AI models in earthquake prediction remains a concern. As noted by Fayaz and Galasso, the "black box" nature of some advanced AI algorithms can make it difficult for seismologists to understand and trust the model's decision-making process [42]. This lack of interpretability can hinder the adoption of AI systems in critical early warning applications. Future research directions should focus on addressing these challenges and advancing the field of AI-driven earthquake prediction. One promising area is the development of physics-informed machine learning models. As suggested by Karpatne et al., integrating domain knowledge from earthquake physics into AI models can improve their accuracy and interpretability, potentially leading to more reliable prediction systems [43]. Improving data collection methods and enhancing data quality is another crucial area for future work. Barnes et al. propose the use of IoT devices integrated with tools like Circulation Obviation Retrofit Kits (CORKs) to enable real-time transmission of hydrogeological measurements influencing earthquakes [44]. This approach could provide the high-resolution data needed to train more accurate AI models. Enhancing the generalizability of AI models across different seismic regions is also a key research direction. Huang et al. demonstrate the potential of convolutional neural networks in detecting and classifying seismic events across various geographical areas [45]. This approach could be particularly valuable for regions with sparse seismic networks or limited historical data. Lastly, addressing ethical considerations and societal impacts of AI-driven earthquake prediction systems is crucial. As highlighted by Pugliese et al., the deployment of AI in critical applications like disaster management raises important questions about transparency, accountability, and public trust [46]. Future research should consider these aspects alongside technical developments. Hence, while AI techniques offer significant improvements in earthquake prediction and early warning, addressing the challenges of data quality, model interpretability, and operational integration is crucial for realizing their full potential. Collaborative efforts between seismologists, AI researchers, and policymakers will be essential to advance these technologies and enhance our ability to mitigate the impacts of seismic events.

3 AI Techniques for Flood Prediction and Management

Flooding is one of the most frequent and damaging natural disasters, affecting millions of people worldwide each year. The complexity of flood prediction and management necessitates the use of advanced technologies to improve accuracy and response times. AI has become an essential tool in flood prediction and management, offering innovative methods to forecast floods, assess risks, and optimize emergency response. Figure 2 show chart of how AI can be used in the flood prediction.

3.1 Machine Learning Models for Flood Forecasting

Machine learning (ML) models have emerged as powerful tools for flood forecasting, offering improved accuracy and adaptability compared to traditional methods. Various ML techniques have been applied to predict flood events, water levels, and inundation areas, leveraging the ability of these models to process large amounts of diverse data and capture complex, non-linear relationships. One of the most widely used ML models for flood forecasting is the Artificial Neural Network (ANN). Mosavi et al. conducted a comprehensive review of ANN applications in flood forecasting, highlighting their effectiveness in predicting river flow and water levels [47]. ANNs have shown particular promise in short-term flood forecasting, with some studies reporting improved accuracy compared to traditional hydrological models. Support Vector Machines (SVM) have also been successfully applied to flood prediction. Tehrany et al. demonstrated the use of SVM for flood susceptibility mapping, achieving high accuracy in identifying flood-prone areas [48]. The ability of SVM to handle high-dimensional data makes it particularly useful for integrating multiple flood-related factors. Random Forest (RF) is another ML algorithm that has shown promise in flood forecasting. Munoz et al. used RF for real-time flood forecasting, demonstrating its ability to handle non-linear relationships and its robustness to noise in the input data [49]. RF has also been effective in feature selection, helping identify the most important variables for flood prediction.

Deep learning models, particularly Long Short-Term Memory (LSTM) networks, have gained popularity in recent years due to their ability to capture long-term

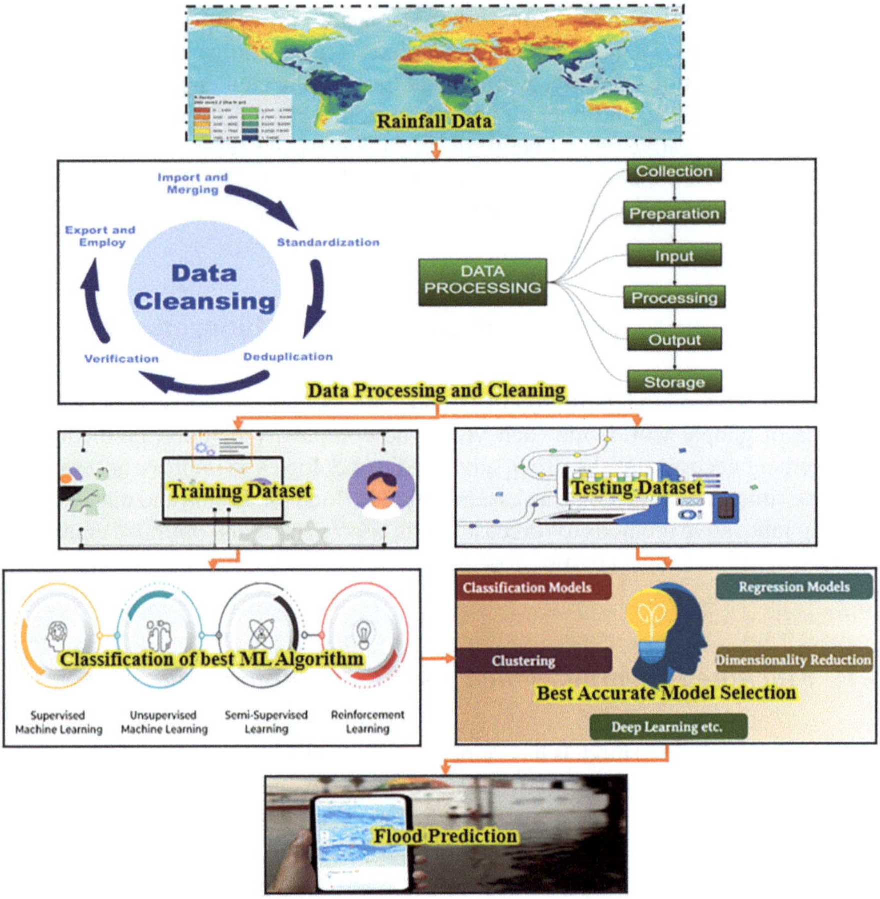

Fig. 2 Flow chart for flood prediction with the help of artificial intelligence

dependencies in time series data. Kratzert et al. applied LSTM networks to rainfall-runoff modelling, achieving state-of-the-art performance in predicting streamflow [50]. Google's operational flood forecasting system, as described by Nevo et al., uses LSTM networks for stage forecasting, demonstrating their effectiveness in real-world applications [51]. Ensemble methods, which combine multiple ML models, have also shown promise in flood forecasting. Bui et al. developed an ensemble model combining SVM, RF, and Logistic Regression for flood susceptibility mapping, achieving higher accuracy than individual models [52]. More recently, Convolutional Neural Networks (CNNs) have been applied to flood forecasting, particularly for processing spatial data such as satellite imagery. Peng et al. used CNNs for rapid flood mapping from satellite images, demonstrating their potential for real-time flood monitoring and response [53]. While these ML models have shown significant promise, challenges remain in their application to flood forecasting. These include the need for large amounts of high-quality training data, the interpretability of complex models, and the generalizability of models across different geographical regions. Ongoing research is focused on addressing these challenges and developing hybrid models that combine the strengths of different ML approaches with physical understanding of flood processes.

3.2 Remote Sensing and AI Integration

The integration of Artificial Intelligence (AI) with remote sensing technologies has revolutionized flood prediction and management capabilities. This synergy allows for the processing and analysis of vast amounts of spatial and temporal data, providing more accurate and timely flood forecasts. Remote sensing technologies, including satellite imagery, aerial photography, and LiDAR (Light Detection and Ranging), combined with AI algorithms, offer unprecedented insights into flood dynamics and risk assessment. Satellite-based remote sensing has emerged as a crucial tool in flood monitoring and prediction. Optical sensors on satellites provide high-resolution imagery that AI algorithms can analyze to detect changes in land use, water bodies, and vegetation cover—all critical factors in flood risk assessment. Synthetic Aperture Radar (SAR) technology has proven particularly valuable due to its ability to penetrate cloud cover and operate in day or night conditions. Huang et al. demonstrated the effectiveness of combining SAR data with deep learning algorithms for rapid and accurate flood mapping, achieving over 90% accuracy in delineating flood extents [54]. The integration of multi-sensor data has further enhanced flood prediction capabilities. Peng et al. developed a novel approach combining optical and SAR satellite data with deep learning models to improve flood inundation mapping [53]. Their method showed significant improvements in accuracy and timeliness compared to traditional remote sensing techniques.

AI algorithms, particularly deep learning models, have shown remarkable ability in extracting meaningful patterns from complex remote sensing data. Convolutional Neural Networks (CNNs) have been widely applied in flood extent mapping and damage assessment. For instance, Nemni et al. utilized a CNN-based approach to analyze multi-temporal satellite imagery for rapid flood detection, demonstrating its potential for near real-time flood monitoring [55]. The integration of remote sensing and AI has also improved the accuracy of hydrological models used in flood forecasting. Sit et al. proposed a deep learning framework that incorporates remote sensing data to enhance the performance of traditional hydrological models, resulting in more accurate flood predictions, especially in data-scarce regions [56]. Unmanned Aerial Vehicles (UAVs) or drones equipped with various sensors have become valuable tools in flood management. AI algorithms can process drone-captured imagery to provide high-resolution, real-time data on flood extents and impacts. Feng et al. developed an AI-powered drone system for rapid flood mapping and damage assessment, showcasing its potential for supporting emergency response efforts [57]. The integration of AI with remote sensing has also facilitated the development of early warning systems. Saravi et al. proposed an AI-based framework that combines satellite data with ground-based sensors for real-time flood forecasting and warning [58]. Their system demonstrated improved lead times and accuracy in predicting flood events. Despite these advancements, challenges remain in the integration of remote sensing and AI for flood prediction. These include the need for high-quality training data, the complexity of interpreting AI model outputs, and the computational resources required for processing large volumes of remote sensing data. Additionally, there is a growing need to address issues of data privacy and security in the use of high-resolution remote sensing data. Future research directions in this field include the development of more robust AI models that can handle the heterogeneity and uncertainty inherent in remote sensing data, the integration of diverse data sources including social media and IoT devices, and the advancement of explainable AI techniques to improve the interpretability of flood prediction models. Hence, the integration of remote sensing and AI presents a powerful approach to flood prediction and management. As these technologies continue to evolve, they promise to provide more accurate, timely, and actionable information for flood risk assessment and mitigation strategies.

3.3 Real-Time Flood Monitoring and Early Warning Systems

Real-time flood monitoring and early warning systems powered by artificial intelligence have emerged as crucial tools in mitigating the impacts of flood events. These systems integrate various data sources, including sensor networks, weather stations, satellite imagery, and social media, to provide timely and accurate flood predictions and warnings. One of the key advantages of AI-powered systems is their ability to process and analyze large volumes of heterogeneous data in real-time. Mosavi et al. conducted a comprehensive review of machine learning models for flood prediction, highlighting the potential of AI techniques in improving the accuracy and lead time of flood forecasts [47]. Their study emphasized the importance of real-time data processing in enhancing the performance of early warning systems. The integration of Internet of Things (IoT) devices with AI algorithms has significantly improved the capabilities of flood monitoring systems. Perera et al. proposed an IoT-based flood monitoring and early warning system that utilizes machine learning algorithms to process data from various sensors [59]. Their system demonstrated improved accuracy in predicting flood events and provided timely warnings to affected communities. Satellite-based remote sensing data has also played a crucial role in enhancing flood monitoring and prediction capabilities. Shen et al. developed a deep learning approach for rapid flood mapping using multi-temporal satellite imagery [60]. Their method showed promising results in quickly identifying flood-affected areas, which is essential for timely response and resource allocation during flood events.

Social media data has emerged as a valuable source of real-time information during flood events. Wang et al. proposed a deep learning framework for flood detection and mapping using social media images [61]. Their approach demonstrated the potential of leveraging user-generated content to complement traditional flood monitoring techniques and improve situational awareness during flood events. The integration of multiple data sources and AI techniques has led to the development of more comprehensive and robust early warning systems. Khosravi et al. presented an integrated flood susceptibility mapping approach that combines machine learning algorithms with GIS and remote sensing data [62]. Their system showed improved performance in identifying flood-prone areas and could potentially be adapted for real-time flood warning applications. Despite the advancements in AI-powered flood monitoring and early warning systems, challenges remain in their widespread adoption and implementation. These include issues related to data quality and availability, model interpretability, and the need for localized calibration of AI models to account for regional variations in flood dynamics. Future research directions in this field include the development of more robust and adaptable AI models that can handle the uncertainties associated with climate change and land-use modifications. Additionally, there is a growing emphasis on developing explainable AI techniques to enhance trust and understanding in the predictions made by these systems. Hence, AI-powered real-time flood monitoring and early warning systems represent a significant advancement in flood management capabilities. By leveraging diverse data sources and advanced machine learning techniques, these systems offer the potential to provide more accurate, timely, and actionable flood warnings, ultimately contributing to improved flood resilience and reduced impacts on communities.

3.4 Flood Risk Assessment and Management

Artificial Intelligence (AI) has emerged as a powerful tool in flood risk assessment and management, offering enhanced capabilities to analyze complex data sets, identify vulnerable areas, and develop more effective mitigation strategies. This subsection reviews recent advancements in AI applications for flood risk assessment and management, highlighting key methodologies and their impacts on improving flood resilience. AI-driven flood risk assessment models have demonstrated significant improvements in accuracy and spatial resolution compared to traditional methods. Tehrany et al. employed a novel ensemble model combining Support Vector Machine (SVM) and Random Forest (RF) algorithms to create flood susceptibility maps [63]. Their approach integrated various geomorphological and hydrological factors, achieving high accuracy in identifying flood-prone areas and outperforming individual machine learning models. Deep learning techniques have shown particular promise in flood risk assessment. Khosravi et al. utilized a deep neural network approach to develop flood susceptibility maps, incorporating a wide range of environmental variables [64]. Their model demonstrated superior performance compared to conventional machine learning methods, especially in capturing complex non-linear relationships between flood occurrence and contributing factors. The integration of AI with remote sensing data has greatly enhanced flood risk assessment capabilities.

Pham et al. developed a novel approach combining Convolutional Neural Networks (CNNs) with multi-temporal satellite imagery to assess flood risk [65]. Their method not only accurately identified flood-prone areas but also captured temporal changes in flood susceptibility, providing valuable insights for long-term flood management strategies.

AI techniques have also been applied to improve the assessment of flood impacts on infrastructure and communities. Wagenaar et al. used machine learning algorithms to predict flood damage to buildings, incorporating factors such as flood depth, duration, and building characteristics [66]. Their model showed significant improvements in damage estimation accuracy compared to traditional stage-damage curves, offering valuable tools for urban planners and insurance companies. In the realm of flood risk management, AI has facilitated the development of more adaptive and efficient strategies. Zischg et al. employed machine learning techniques to optimize flood protection measures, considering multiple objectives such as cost-effectiveness and environmental impact [67]. Their approach demonstrated the potential of AI in supporting complex decision-making processes in flood risk management. Real-time flood risk assessment has been another area where AI has shown considerable promise. Sit et al. developed a deep learning framework for real-time flood inundation mapping using social media data and remote sensing imagery [68]. Their approach enabled rapid assessment of flood extents during events, providing crucial information for emergency responders and affected communities. Despite these advancements, challenges remain in the widespread adoption of AI for flood risk assessment and management. Issues such as data quality and availability, model interpretability, and the need for local calibration of AI models continue to be areas of active research. Additionally, there is a growing emphasis on developing explainable AI techniques to enhance trust and understanding in the predictions made by these systems, particularly when used for critical decision-making in flood management. Future research directions in this field include the development of more robust and adaptable AI models that can account for changing climate patterns and land-use dynamics. There is also a need for improved integration of AI-driven risk assessment tools with existing flood management frameworks and decision-support systems. To summarise, AI has demonstrated significant potential in enhancing flood risk assessment and management practices. By leveraging advanced machine learning techniques and diverse data sources, AI-driven approaches offer more accurate, timely, and comprehensive insights into flood risks. As these technologies continue to evolve, they promise to play an increasingly important role in building flood-resilient communities and infrastructure.

3.5 Post Flood Damage Assessment and Recovery

AI has emerged as a powerful tool in post-flood damage assessment and recovery management, offering rapid, accurate, and comprehensive analysis capabilities that significantly enhance traditional methods. This subsection reviews recent advancements in AI applications for post-flood scenarios, highlighting key methodologies and their impacts on improving recovery efforts. One of the most significant applications of AI in post-flood assessment is the analysis of satellite and aerial imagery. Amit and Aoki developed a deep learning model using convolutional neural networks (CNNs) to automatically detect and classify flood damage from high-resolution satellite images [69]. Their approach demonstrated high accuracy in identifying various levels of structural damage, enabling rapid assessment of large areas and prioritization of recovery efforts. Building on this, Sublime and Kalinicheva, proposed an integrated framework combining multi-temporal satellite imagery with social media data for comprehensive flood damage assessment [70]. Their machine learning model not only identified physical damage but also assessed the social and economic impacts of flooding, providing a more holistic view of disaster effects and informing targeted recovery strategies. AI-driven damage assessment has also been extended to critical infrastructure. Kharazi and Behzadan developed a deep learning model for assessing flood damage to road networks using a combination of satellite imagery and street-level photographs [71]. Their approach enabled rapid identification of damaged road segments, facilitating efficient allocation of repair resources and restoration of transportation networks.

AI has also shown promise in optimizing resource allocation during the recovery phase. Nojavan et al. developed a machine learning algorithm for dynamic resource allocation in post-flood scenarios [72]. Their model adaptively learned from ongoing recovery efforts to optimize the distribution of personnel, equipment, and supplies across affected areas, significantly improving the efficiency of recovery operations. Furthermore, AI has been leveraged to enhance community engagement in the recovery process. Nguyen et al. created an AI-powered platform that analysed social media posts and community feedback to identify unmet needs and emerging issues during flood recovery [73]. This approach enabled more responsive and community-centered recovery efforts, ensuring that resources were directed to areas of greatest need. Despite these advancements, challenges remain in the widespread adoption of AI for post-flood damage assessment and recovery. Issues such as data quality and availability, model interpretability, and the need for local calibration of AI models continue to be areas of active research. Additionally, there is a growing emphasis on developing explainable AI techniques to enhance trust and understanding in the predictions made by these systems, particularly when used for critical decision-making in disaster recovery contexts. Future research directions in this field include the development of more robust and adaptable AI models that can account for the complex and dynamic nature of flood impacts. There is also a need for improved integration of AI-driven assessment tools with existing disaster management frameworks and decision-support systems.

4 AI Applications in Hurricane and Typhoon Management

Hurricanes and typhoons are among the most destructive natural disasters, characterized by their intense winds, heavy rainfall, and storm surges. Effective management of these events requires accurate forecasting, timely warnings, and efficient response strategies. Figure 3 shows the integration of AI in hurricane and typhoon management. AI has emerged as a powerful tool in hurricane and typhoon management, offering innovative solutions to improve prediction, monitoring, and response efforts.

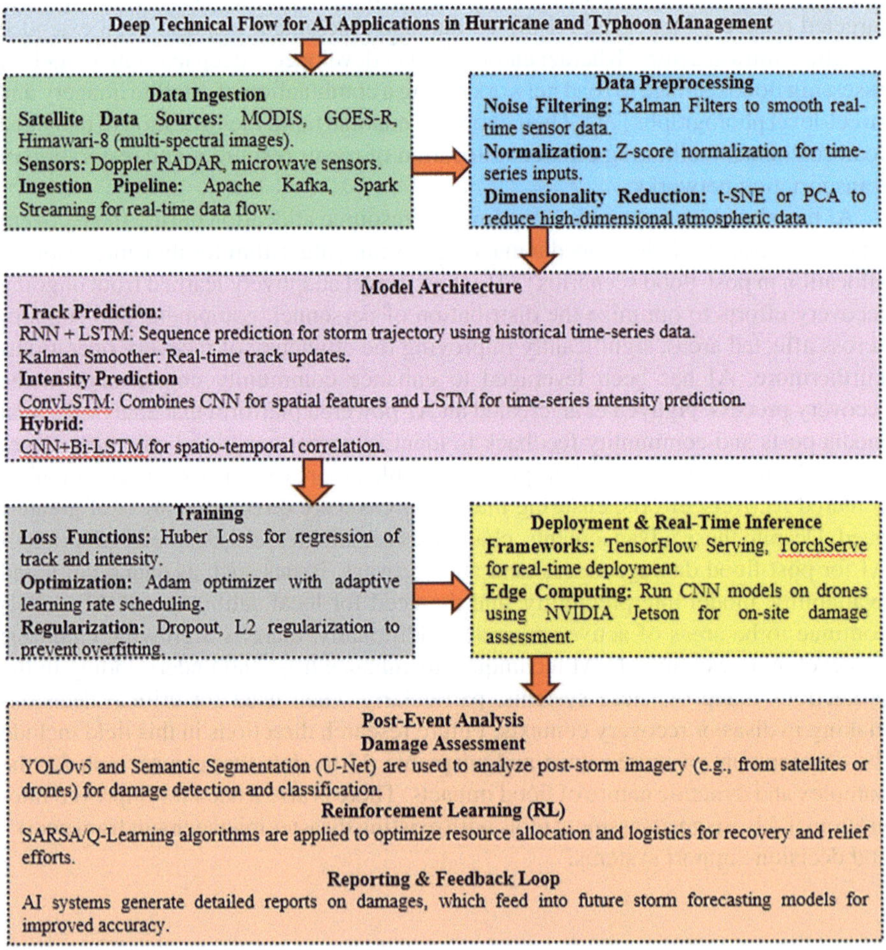

Fig. 3 Flow chart on integration of AI in hurricane and typhoon management

4.1 Enhanced Hurricane and Typhoon Forecasting with AI

AI has revolutionized hurricane and typhoon forecasting, offering significant improvements in accuracy, lead time, and detailed predictions. This subsection reviews recent advancements in AI-driven forecasting for tropical cyclones, highlighting key methodologies and their impacts on improving storm prediction and management. Machine learning models have demonstrated remarkable capabilities in analysing complex meteorological data to predict hurricane trajectories and intensities. Boussioux et al. introduced a novel multimodal machine learning framework called Hurricast for both intensity and track forecasting [74]. Their approach combines gradient-boosted trees with deep learning architectures, including CNNs, GRUs, and Transformers, to process diverse data sources such as historical storm data and reanalysis atmospheric images. The framework showed competitive performance compared to operational forecasts in the North Atlantic and Eastern Pacific basins, highlighting the potential of AI in operational hurricane forecasting. Deep learning models have shown particular promise in extracting meaningful patterns from complex meteorological data. Chen et al. developed a CNN approach for tropical cyclone track prediction using multi-channel atmospheric images [75]. Their model demonstrated improved accuracy in predicting hurricane trajectories compared to traditional numerical weather prediction models, especially for longer forecast lead times. The integration of multiple data sources has been key to these advancements. Giffard-Roisin et al. proposed a CNN-based model that combines historical trajectory data with reanalysis atmospheric fields to predict tropical cyclone intensity and track simultaneously [76]. Their approach showed significant improvements over baseline methods, particularly in predicting rapid intensification events, which are notoriously difficult to forecast.

AI models have also shown promise in addressing specific challenges in hurricane forecasting. For instance, Mudigonda et al. developed a semi-supervised deep learning model for tropical cyclone intensity estimation using satellite imagery [77]. Their approach demonstrated the ability to accurately estimate cyclone intensity even with limited labelled data, addressing a common challenge in tropical cyclone research. Despite these advancements, challenges remain in the widespread adoption of AI for hurricane forecasting. Issues of model interpretability, the need for extensive high-quality training data, and the integration of AI predictions with existing forecasting systems continue to be areas of active research. Future research directions in this field include the development of more robust and adaptable AI models that can handle the uncertainties associated with climate change, the integration of AI forecasts with traditional numerical weather prediction models, and the advancement of explainable AI techniques to enhance trust and understanding in AI-generated forecasts. Hence, AI-enhanced forecasting for hurricanes and typhoons represents a significant advancement in meteorological science and disaster management. These models offer the potential to provide more accurate, timely, and detailed predictions, ultimately contributing to improved preparedness and response strategies for tropical cyclones.

4.2 Satellite Data Analysis

The integration of AI with satellite technology has revolutionized hurricane and typhoon monitoring, offering unprecedented capabilities in storm detection, tracking, and intensity estimation. This subsection reviews recent advancements in AI-enhanced satellite data analysis for hurricane and typhoon management, highlighting key methodologies and their impacts on improving forecasting and early warning systems. One of the most significant applications of AI in satellite data analysis is the automated detection and tracking of tropical cyclones. Kumler-Bonfanti et al. developed a deep learning model using CNNs to identify and track tropical cyclones from satellite imagery [78]. Their approach demonstrated high accuracy in detecting cyclone centres and estimating their intensities, outperforming traditional methods in both speed and precision. AI techniques have also shown promise in estimating hurricane intensities from satellite data. Lee et al. proposed a deep learning framework that combines multiple satellite channels to estimate tropical cyclone intensity [79]. Their model achieved comparable accuracy to human expert estimates while significantly reducing processing time, potentially enabling more frequent and timely intensity updates during operational forecasting. The use of AI in analyzing multi-spectral satellite imagery has led to improved understanding of hurricane structure and evolution. Wimmers et al. developed an AI-based system that uses deep learning to analyze infrared satellite imagery and estimate tropical cyclone intensity [80]. Their approach showed particular skill in identifying rapid intensification events, which are traditionally challenging to forecast.

AI has also been applied to enhance the resolution and quality of satellite imagery used in hurricane monitoring. Li et al. presented a super-resolution technique using generative adversarial networks (GANs) to improve the spatial resolution of satellite-based precipitation estimates in tropical cyclones [81]. This approach could potentially lead to more detailed and accurate representations of hurricane rainfall patterns. Furthermore, AI techniques have been employed to fuse data from multiple satellite sensors, providing a more comprehensive view of hurricane development. Leroux et al. demonstrated the use of a deep learning model to integrate data from microwave and infrared sensors, improving the accuracy of tropical cyclone centre fixes and intensity estimates, especially in cases where traditional methods struggle due to obscured eye features [82]. The application of AI to historical satellite data has also yielded valuable insights. Kordmahalleh et al. used a recurrent neural network (RNN) approach to predict hurricane trajectories based on historical data [83]. Their model demonstrated improved accuracy in predicting hurricane paths compared to traditional statistical methods, particularly for longer forecast horizons. This approach has the potential to improve our understanding of long-term trends in hurricane activity and their relationship to climate change. Despite these advancements, challenges remain in the widespread adoption of AI for satellite-based hurricane analysis. Issues such as model interpretability, the need for large training datasets, and the integration of AI-derived insights with existing forecasting systems continue to be areas of active research. Future research directions in this field include the development of more robust and adaptable AI models that can handle the diverse range of tropical cyclone structures and environments, the integration of AI-based satellite analysis with numerical weather prediction models, and the advancement of explainable AI techniques to enhance trust and understanding in AI-derived hurricane analyses. Hence, AI-enhanced satellite data analysis represents a significant advancement in hurricane and typhoon monitoring and forecasting. By leveraging advanced machine learning techniques and the wealth of available satellite data, these approaches offer the potential to provide more accurate, timely, and comprehensive insights into tropical cyclone behaviour, ultimately contributing to improved early warning systems and disaster preparedness strategies.

4.3 Real-Time Storm Tracking and Prediction

AI-powered real-time tracking systems have revolutionized the monitoring and prediction of hurricanes and typhoons, offering enhanced capabilities for tracking these storms as they develop and move. These systems leverage artificial intelligence to analyze vast amounts of data from diverse sources, including weather stations, satellites, ocean buoys, and even social media, to provide up-to-date information on storm conditions with unprecedented speed and accuracy. One of the key advantages of AI in real-time storm tracking is its ability to process and interpret data much faster than traditional methods. Chen et al. demonstrated the effectiveness of deep learning models in rapidly analyzing satellite imagery for tropical cyclone intensity estimation [75]. Their convolutional neural network approach achieved comparable accuracy to human expert estimates while significantly reducing processing time, enabling more frequent and timely intensity updates during operational forecasting. The integration of multiple data sources has been crucial in improving the accuracy of real-time storm tracking. Giffard-Roisin et al. proposed a novel deep learning framework that combines historical trajectory data with reanalysis atmospheric fields to predict tropical cyclone intensity and track simultaneously [76]. This approach showed significant improvements over baseline methods, particularly in predicting rapid intensification events, which are notoriously difficult to forecast in real-time.

AI algorithms have also shown promise in extracting valuable information from non-traditional data sources. For instance, Kordmahalleh et al. developed a recurrent neural network model to predict hurricane trajectories based on historical data and real-time observations [83]. Their approach demonstrated improved accuracy in predicting hurricane paths compared to traditional statistical methods, especially for longer forecast horizons. The application of AI to real-time storm tracking extends beyond just prediction. Chattopadhyay et al. introduced a deep learning approach for automated tropical cyclone intensity estimation using satellite imagery [84]. Their model not only provided accurate intensity estimates but also generated attention maps highlighting the most relevant features for the estimation, offering insights into the decision-making process of the AI system. Despite these advancements, challenges remain in the widespread adoption of AI for real-time storm tracking and prediction. Issues such as model interpretability, the need for large amounts of high-quality training data, and the integration of AI predictions with existing forecasting systems continue to be areas of active research. Future research directions in this field include the development of more robust and adaptable AI models that can handle the diverse range of tropical cyclone structures and environments, the integration of AI-based tracking systems with numerical weather prediction models, and the advancement of explainable AI techniques to enhance trust and understanding in AI-derived storm analyses. Hence, AI-powered real-time storm tracking and prediction systems represent a significant advancement in hurricane and typhoon management. By leveraging advanced machine learning techniques and diverse data sources, these approaches offer the potential to provide more accurate, timely, and comprehensive insights into tropical cyclone behaviour, ultimately contributing to improved early warning systems and disaster preparedness strategies.

4.4 Risk Assessment and Impact Prediction

AI has emerged as a powerful tool in assessing the potential impact of hurricanes and typhoons on communities and infrastructure. By leveraging machine learning algorithms to analyze vast amounts of historical data, geographic information, and real-time storm conditions, AI models can provide more accurate and timely predictions of storm impacts, enabling better preparedness and response strategies. One of the key areas where AI has shown significant promise is in damage estimation. Soni et al. developed a deep learning model that combines satellite imagery with socioeconomic data to predict hurricane-induced economic losses [85]. Their model demonstrated superior performance compared to traditional statistical methods, particularly in capturing the spatial heterogeneity of damage patterns. AI has also been applied to improve evacuation planning. In the realm of evacuation planning, AI has shown promise in optimizing strategies. Li et al. analysed empirical evacuation response curves during Hurricane Irene in Cape May County, New Jersey [86]. While this study did not use AI directly, it provides valuable data that could be used to train and validate AI models for evacuation planning. The empirical data on evacuation behaviour could inform machine learning approaches to optimize evacuation routes and timing.

AI is also being applied to predict storm surge, a major cause of damage during hurricanes. Tadesse et al. developed a machine learning model combining numerical simulations with artificial neural networks to provide rapid and accurate predictions of coastal flooding [87]. Their approach significantly reduced computational time while maintaining high accuracy, potentially enabling more timely warnings for coastal communities. Furthermore, AI models are being used to assess the impacts of hurricanes on infrastructure and communities. Wang et al. proposed a deep learning framework to analyze post-hurricane satellite imagery for rapid damage assessment [88]. Their model demonstrated high accuracy in identifying different levels of structural damage, potentially expediting post-disaster response efforts and informing long-term resilience strategies. Despite these advancements, challenges remain in the widespread adoption of AI for hurricane impact prediction. Issues such as data quality, model interpretability, and the need to account for climate change effects on historical patterns continue to be areas of active research. Additionally, there is a growing emphasis on developing AI models that can incorporate real-time data streams during an ongoing event to provide continuously updated impact assessments. Hence, AI-driven risk assessment and impact prediction for hurricanes and typhoons represent a significant advancement in disaster management capabilities. These approaches offer the potential to provide more accurate, timely, and comprehensive insights into storm impacts, ultimately contributing to improved preparedness, response, and resilience strategies for vulnerable communities.

4.5 Post-Storm Damage Assessment and Recovery

In the aftermath of hurricanes and typhoons, rapid and accurate damage assessment is crucial for effective recovery efforts. Artificial Intelligence (AI) has emerged as a powerful tool in this domain, offering capabilities that significantly enhance the speed and precision of post-storm evaluations and recovery planning. One of the primary applications of AI in post-storm assessment is the analysis of satellite and aerial imagery. Doshi et al. developed a convolutional neural network (CNN) approach for rapid damage assessment using post-hurricane satellite imagery [89]. Their model demonstrated high accuracy in identifying different levels of building damage, potentially expediting post-disaster response efforts. AI-driven damage assessment is not limited to building structures. Rahnemoonfar et al. proposed a deep learning framework for flood extent mapping from satellite imagery in the aftermath of hurricanes [90]. Their method accurately identified flood-affected areas, providing valuable insights for both immediate response and long-term resilience planning. The integration of multiple data sources has proven crucial in enhancing the accuracy and comprehensiveness of post-storm assessments. Hao and Wang developed a machine learning approach that combines remote sensing data with social media information to assess hurricane damage [91]. This holistic approach enables a more nuanced understanding of disaster impacts, considering both physical damage and social impacts.

AI technologies have also shown promise in optimizing recovery and reconstruction efforts. Nojavan et al. proposed a machine learning-based framework for post-disaster resource allocation [72]. Their model adaptively learned from ongoing recovery efforts to optimize the distribution of personnel, equipment, and supplies across affected areas, potentially improving the efficiency of recovery operations. Despite these advancements, challenges remain in the widespread adoption of AI for post-storm assessment and recovery. Issues such as data quality, model interpretability, and the need to account for local contextual factors continue to be areas of active research. Hence, AI-driven approaches for post-storm damage assessment and recovery represent a significant advancement in disaster management capabilities. These technologies offer the potential to provide more accurate, timely, and comprehensive insights into storm impacts, ultimately contributing to more effective and efficient recovery strategies.

5 AI in Managing Volcanic Eruptions

Volcanic eruptions, while less frequent than other natural disasters, can have catastrophic consequences for surrounding regions. Effective management of volcanic eruptions involves accurate prediction, real-time monitoring, and efficient response strategies. Figure 4 shows the flow chart on how AI can be used in the prediction of Volcanic Eruptions. AI technologies have emerged as valuable tools in enhancing volcanic eruption management by improving prediction accuracy, monitoring capabilities, and response efforts.

5.1 Predictive Models for Volcanic Eruptions

The application of artificial intelligence and machine learning techniques to volcanic eruption prediction has revolutionized the field of volcanology in recent years. These advanced computational methods have significantly enhanced our ability to analyze complex, multivariate datasets and identify subtle precursors of volcanic activity that may elude traditional analysis methods. One of the most promising applications of AI in volcanic eruption prediction is the integration of multiple data sources. Ren et al. developed a novel deep learning framework that combines seismic, geodetic, and gas emission data to forecast volcanic unrest [92]. Their model demonstrated superior performance in detecting precursory signals compared to conventional statistical methods, potentially extending the warning time for impending eruptions. The use of CNNs has shown particular promise in analyzing satellite imagery for signs of volcanic unrest. Valade et al. applied a CNN-based approach to multispectral satellite images to detect and classify volcanic plumes [93]. Their method achieved high accuracy in distinguishing between ash-rich and ash-poor plumes, providing crucial information for aviation safety and hazard assessment. RNNs have been effectively employed to capture the temporal dynamics of volcanic systems. Titos et al. utilized LSTM networks to analyze time series of seismic events preceding eruptions [94]. Their model successfully identified patterns in seismic activity that were indicative of imminent eruptions, outperforming traditional time series analysis methods.

The integration of machine learning with physics-based models has also shown great potential. Parra describes a volcanic ash forecasting system for Ecuador that utilizes coupled Eulerian numerical models, specifically WRF for meteorology and FALL3D for ash dispersion [95]. This system generates forecasts for the Tungurahua and Cotopaxi volcanoes, using eruption source parameters derived from past

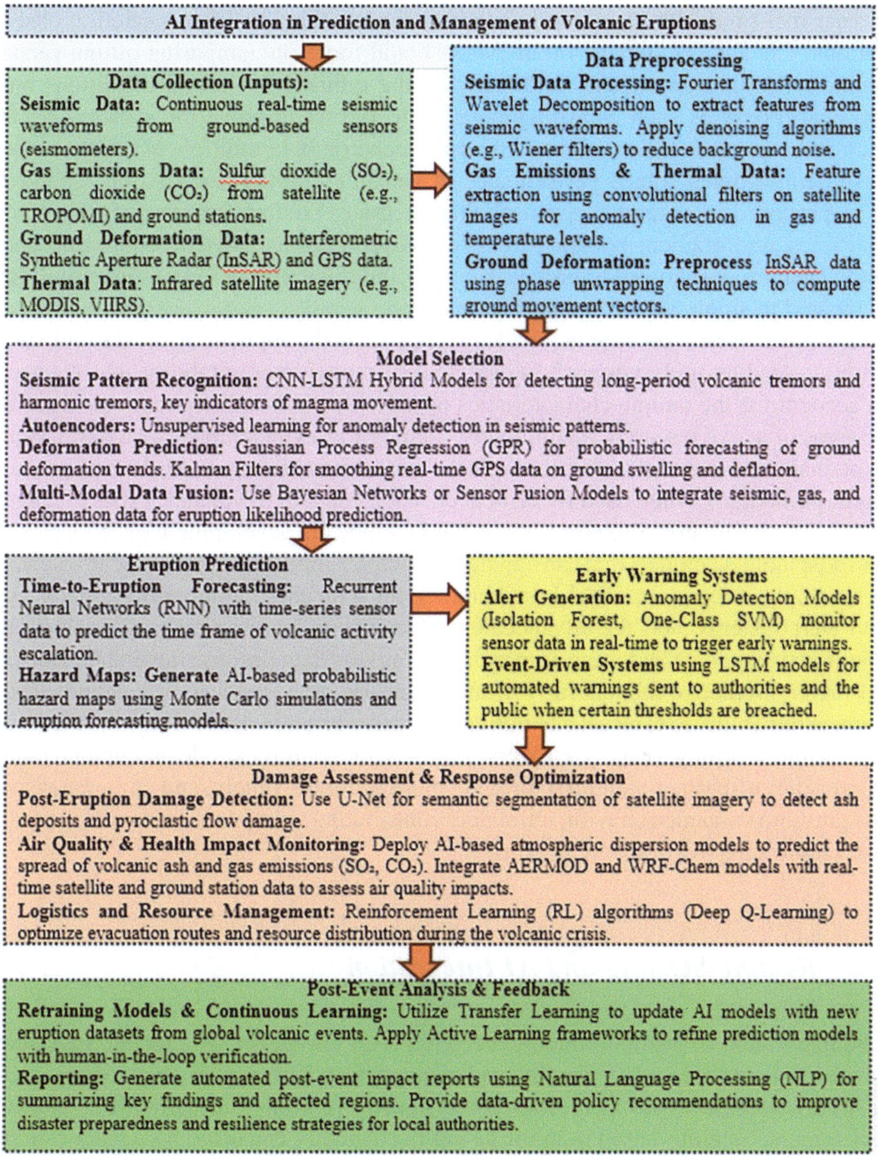

Fig. 4 Flow chart for how AI integrates in prediction and management of volcanic eruptions

events in 2013 and 2015 respectively. The system employs an offline approach, first forecasting meteorology and then ash dispersion separately, which allows for the reuse of meteorological data but does not account for ash-atmosphere interactions. Once an eruption begins, the system can provide high spatial resolution forecasts in near real-time, although cloud cover can limit validation against satellite observations. The authors suggest future work should focus on comparing offline versus online modelling approaches and studying the influence of initial conditions and physical processes such as ash aggregation to further improve the system's accuracy and capabilities. Ensemble methods, which combine multiple ML algorithms, have demonstrated robust performance in volcanic eruption forecasting. Cannavò et al. proposed an ensemble approach that integrates random forests, support vector machines, and gradient boosting trees to analyze multiparametric monitoring data [96]. Their model showed improved accuracy and reliability in forecasting eruptions compared to single-algorithm approaches.

Despite these advancements, challenges remain in the application of AI to volcanic eruption prediction. Issues such as data quality, model interpretability, and the need to account for the unique characteristics of individual volcanic systems continue to be areas of active research. Additionally, there is a growing emphasis on developing explainable AI techniques to enhance trust and understanding in AI-derived predictions, particularly when used for critical decision-making in volcanic crisis management. Future research directions in this field include the development of more robust and adaptable AI models that can handle the complex, non-linear behavior of volcanic systems, the integration of AI-based predictions with existing monitoring networks and alert systems, and the advancement of real-time learning algorithms that can continuously update and improve predictions as new data becomes available. Hence, AI-driven predictive models for volcanic eruptions represent a significant advancement in our ability to forecast and mitigate volcanic hazards. By leveraging the power of machine learning to analyze diverse datasets and identify subtle precursors, these approaches offer the potential to provide more accurate, timely, and comprehensive insights into volcanic behavior, ultimately contributing to improved risk management and public safety in volcanic regions.

5.2 Remote Sensing and AI Integration

The integration of artificial intelligence (AI) with remote sensing technologies has revolutionized volcanic monitoring and management. This synergy has led to significant advancements in early warning systems, eruption forecasting, and hazard assessment. Satellite-based monitoring has been greatly enhanced by AI algorithms, enabling the detection of subtle changes in volcanic systems. The MOUNTS (Monitoring Unrest from Space) project, developed by Valade et al., demonstrates the power of combining multiple satellite sensors with AI for comprehensive volcanic monitoring. This system integrates data from Sentinel-1, Sentinel-2, and Sentinel-5P satellites to track various volcanic parameters, including surface deformation, thermal anomalies, and gas emissions [93]. Machine learning techniques have proven particularly effective in processing vast amounts of satellite imagery. Anantrasirichai et al., developed a convolutional neural network (CNN) approach for automatically detecting volcanic deformation in InSAR images [97]. Their method significantly reduced the time required for analysis and improved the accuracy of deformation detection, potentially enabling near-real-time monitoring of ground displacements associated with volcanic activity [97].

The integration of multiple data sources has further enhanced our ability to forecast eruptions. Gaddes et al. demonstrated the potential of combining InSAR time series with machine learning algorithms to identify precursory deformation patterns [98].

Their approach showed promise in distinguishing between different types of volcanic unrest and improving eruption forecasting capabilities. AI-driven analysis of thermal imagery has also made significant strides. Aldeghi et al., developed a deep learning model for automated detection and quantification of lava flows using Sentinel-2 imagery [99]. This method allows for rapid assessment of eruption dynamics and potential hazards, crucial for timely risk management. Despite these advancements, challenges remain in the integration of AI and remote sensing for volcanic monitoring. Issues such as cloud cover, which can obscure satellite imagery, and the need for high-resolution data to detect subtle changes continue to pose difficulties. Additionally, the development of AI models that can generalize across different volcanic systems and types of eruptions remains an active area of research. Future work in this field is likely to focus on improving the resolution and frequency of satellite observations, developing more sophisticated AI algorithms capable of integrating diverse data types, and creating systems for real-time processing and analysis of remote sensing data. As these technologies continue to evolve, their integration promises to significantly enhance our ability to monitor and forecast volcanic eruptions, ultimately contributing to improved hazard management and public safety in volcanic regions.

5.3 Real-Time Monitoring and Early Warning Systems

AI-powered real-time monitoring and early warning systems have become increasingly crucial in managing volcanic eruptions. These systems integrate data from various sensors and sources to provide timely alerts and improve response efforts. Reath et al. developed a machine learning approach to automatically detect volcanic deformation in InSAR data [100]. Their method, using a random forest classifier, achieved an accuracy of 82% in identifying deformation signals, potentially enabling faster detection of pre-eruptive ground movements. Integrating multiple data sources has proven effective in enhancing early warning capabilities. Hebert et al. presented a multi-sensor approach combining satellite-based InSAR, GNSS, and seismic data for near-real-time monitoring of volcanic unrest [101]. Thermal monitoring has also benefited from AI integration. Marchese et al. developed an automated system for detecting and tracking volcanic thermal anomalies using geostationary satellite data and machine learning algorithms. Their approach enables near-real-time monitoring of volcanic hotspots, crucial for identifying new eruptive vents and lava flow paths.

Seismic data analysis remains a cornerstone of early warning systems. Bueno et al. proposed a deep learning approach for real-time classification of volcanic seismic events [102]. Their method, using CNNs, achieved high accuracy in distinguishing between different types of volcanic earthquakes, potentially improving the timeliness of eruption forecasts. Despite these advancements, challenges remain in implementing AI-powered early warning systems. Issues such as false alarms, data quality, and the need for continuous model updating to account for evolving volcanic behaviour are active areas of research. Additionally, the integration of these systems with existing monitoring networks and decision-making processes requires careful consideration. Future developments in this field are likely to focus on improving the integration of diverse data types, enhancing the spatial and temporal resolution of monitoring, and developing more sophisticated AI models capable of capturing the complex dynamics of volcanic systems. As these technologies continue to evolve, they promise to significantly enhance our ability to provide timely and accurate early warnings for volcanic eruptions, ultimately contributing to improved hazard management and public safety in volcanic regions.

5.4 Impact Assessment and Response Optimization

In the aftermath of a volcanic eruption, AI plays a crucial role in assessing impact and optimizing response efforts. Recent advancements in AI-driven analysis of satellite and aerial imagery have significantly enhanced our ability to quickly evaluate affected regions and improve recovery strategies. A notable development in this field is the use of machine learning methods for detecting and retrieving volcanic clouds. For instance, a study by Romeo et al. highlights how machine learning can effectively detect volcanic clouds using multispectral satellite data, providing accurate estimates of volcanic ash distribution and mass loading [103]. This capability is crucial for assessing the immediate impact of eruptions on air quality and aviation safety. Building damage assessment is another crucial aspect of post-eruption impact analysis. Ghaffarian et al. proposed a deep learning-based approach for rapid mapping of building damage caused by natural disasters, including volcanic eruptions [104]. Their method, using a Siamese convolutional neural network, showed promising results in identifying damaged structures from pre- and post-event satellite imagery, enabling faster and more accurate damage assessments. AI is also being applied to optimize evacuation strategies. Jumadi et al. developed an agent-based model integrating geographic information systems (GIS) and machine learning to simulate and evaluate evacuation scenarios for volcanic crises [105]. Their approach considers various factors such as population distribution, road networks, and eruption dynamics to identify optimal evacuation routes and strategies3.

For long-term impact assessment and recovery planning, Deligne et al. utilized an integrated modelling approach to simulate the economic impacts of a volcanic eruption in Auckland, New Zealand [106]. While not explicitly using AI, their work demonstrates the potential for incorporating machine learning techniques to enhance the accuracy and adaptability of such economic impact models. Lastly, Pallister et al. discussed the integration of AI and machine learning in volcano observatories for improved eruption forecasting and risk assessment [107]. They highlighted the potential of AI to process and analyze diverse datasets, including seismic, geodetic, and gas emission data, to provide more comprehensive and timely assessments of volcanic activity and potential impacts. These advancements in AI-driven impact assessment and response optimization offer promising tools for managing volcanic crises. However, challenges remain in refining these models to account for the unique characteristics of different volcanic systems and eruption types. Future research should focus on improving the adaptability of AI models to diverse volcanic scenarios and integrating them more seamlessly with existing disaster response frameworks.

5.5 Risk Assessment and Management

AI models have become indispensable tools in volcanic risk assessment, enabling the analysis of historical eruption data, geological features, and population exposure to predict future eruptions and assess potential impacts on communities and infrastructure. Recent advancements in machine learning have significantly improved the accuracy of eruption forecasts. A study by Salinas et al. demonstrated the use of neural networks to classify volcanic alert states based on seismic data [108]. This approach provides probabilistic forecasts of volcanic activity, enhancing the ability to predict eruptions and manage risks effectively. The integration of AI with remote sensing technologies has also advanced risk assessment capabilities. The MOUNTS system, as discussed by Valade et al., utilizes multisensor satellite data combined with AI algorithms to monitor volcanic activity globally [93]. This system provides real-time assessments of surface deformation, gas emissions, and thermal anomalies, offering critical insights into potential eruption scenarios. AI-driven models are also being used to simulate the socio-economic impacts of volcanic eruptions. Cardwell et al. developed a simulation model to assess land use and economic recovery pathways following eruptions in urban areas [109]. Their model helps identify optimal recovery strategies by simulating different scenarios and their long-term impacts on infrastructure and communities.

In addition to forecasting and impact simulation, AI is enhancing the integration of hazard, exposure, vulnerability, and resilience assessments. The ADVISE model, described by Bonadonna et al., combines these elements to provide comprehensive risk assessments tailored to specific volcanic contexts [110]. This approach allows for both short-term emergency management and long-term risk reduction strategies. Moreover, AI's ability to process large datasets quickly makes it a valuable tool for continuous monitoring and rapid response. A study by Ferrari et al. highlights how AI can be used to analyze multiparametric data from various sensors for near-real-time risk assessment, improving preparedness and response efforts [111]. These advancements demonstrate the potential of AI in enhancing volcanic risk management strategies and disaster preparedness. However, challenges remain in ensuring the adaptability of AI models across different volcanic systems and integrating them seamlessly with existing monitoring frameworks. Future research should focus on refining these models to address diverse volcanic scenarios and improve their practical application in risk management.

6 Future Directions of AI in Natural Disaster Management

The future of AI in natural disaster management is poised to revolutionize how we predict, prepare for, and respond to disasters. Several exciting advancements and integrations are on the horizon, offering the potential to significantly enhance our disaster management capabilities. This section explores key future directions for AI in natural disaster management.

6.1 Integration with Emerging Technologies

6.1.1 Internet of Things (IoT)

The integration of AI with Internet of Things (IoT) devices represents a major advancement in disaster management. IoT devices, such as sensors and smart meters, can provide real-time data on environmental conditions, infrastructure health, and disaster impacts. By combining AI with IoT, we can achieve more comprehensive and timely monitoring.

Smart Sensors: Deploying smart sensors in vulnerable areas allows for continuous monitoring of various environmental parameters such as temperature, humidity, and seismic activity. These sensors can detect early warning signs of natural disasters like floods, earthquakes, or wildfires, providing critical data that AI algorithms can analyze to predict and respond to disasters more effectively [112].

Real-Time Data Collection: IoT devices can collect vast amounts of data in real-time, which AI systems can use to update predictions and manage responses dynamically. For instance, integrating AI with IoT in flood management can improve the accuracy of flood predictions and the efficiency of emergency responses [113–115].

6.1.2 Advanced Data Analytics

The evolution of data analytics and computational power is set to enhance AI's capabilities in disaster management. Advanced analytics can improve the accuracy of AI models, enabling more precise predictions and better-informed decision-making.

Big Data: Leveraging big data technologies enables the processing and analysis of large volumes of disaster-related data from various sources, including social media, satellite imagery, and sensor networks. AI models can utilize this data to identify patterns and trends that might not be apparent from smaller datasets. This approach can enhance the forecasting of natural disasters and optimize resource allocation during crises [116, 117].

Enhanced Computational Power: Advances in computational power, such as those provided by GPUs and cloud computing, allow for the training of more complex AI models and faster processing of large datasets. This improvement will lead to more accurate and timely predictions, as well as better simulations of disaster scenarios [118].

6.2 Collaboration and Knowledge Sharing

6.2.1 Global Collaboration

Global collaboration and knowledge sharing are crucial for advancing AI applications in disaster management. By pooling resources and expertise, countries can develop more effective AI solutions and improve overall disaster resilience.

International Partnerships: Engaging in international partnerships allows for the exchange of data, research findings, and best practices among countries. Collaborative efforts can lead to the development of standardized methods and shared technologies that enhance global disaster management [119, 120].

Global Databases: Establishing global databases of disaster-related data can facilitate more comprehensive analyses and improve AI models. Sharing information on disaster patterns, impacts, and responses can help in creating more accurate prediction models and effective response strategies [121].

6.2.2 Public–Private Partnerships

Public–private partnerships (PPPs) are essential for driving innovation and investment in AI technologies for disaster management. Collaboration between government agencies, businesses, and research institutions can accelerate the development and deployment of AI solutions.

Collaborative Projects: Supporting collaborative projects between various stakeholders can lead to the creation of cutting-edge technologies and applications. For example, partnerships between tech companies and disaster response agencies can result in the development of advanced AI systems for real-time monitoring and response [122].

Investment in Research: Public and private sectors can jointly invest in research and development to explore new AI techniques and applications for disaster management. These investments can foster innovation and bring new solutions to market faster, improving disaster preparedness and response [8, 123].

Hence, the integration of emerging technologies such as IoT and advanced data analytics, along with global and public–private collaboration, represents key future directions for AI in natural disaster management. These advancements hold the potential to significantly enhance our ability to predict, prepare for, and respond to natural disasters, ultimately improving resilience and saving lives.

7 Conclusion

This chapter has explored the transformative potential of AI in predicting and managing natural disasters, focusing on earthquakes, floods, hurricanes, typhoons, and volcanic eruptions. The sources highlight how AI is revolutionizing traditional disaster management practices, leading to significant improvements in accuracy, timeliness, and efficiency across all stages of prediction, monitoring, and response. The ML models have emerged as powerful tools for forecasting various disaster events. For example, Random Forest models have achieved high accuracy in predicting earthquake magnitudes, while deep learning models like CNNs have proven effective in detecting and classifying seismic events, including the differentiation between natural earthquakes and human-induced events. In flood forecasting, ANNs and LSTMs are widely used to predict flood events, water levels, and inundation areas. For hurricane prediction, multimodal machine learning frameworks like Hurricast, which combine gradient-boosted trees with deep learning architectures, have shown promising results in forecasting hurricane trajectories and intensities. The integration of AI with remote sensing technologies has significantly enhanced disaster monitoring capabilities. CNNs have been successfully applied to analyze satellite imagery for rapid damage assessment, flood mapping, and storm tracking. SAR technology combined with deep learning has enabled accurate flood mapping even in challenging conditions like cloud cover. Real-time monitoring and early warning systems powered by AI are crucial for mitigating the impact of disasters. By integrating data from sensor networks, weather stations, satellite imagery, and social media, these systems can provide timely and accurate alerts for floods, volcanic eruptions, and other hazards. The sources also emphasize the challenges and future directions for AI in disaster management. Data quality and availability, model interpretability, and the need for localized calibration of AI models remain key areas for ongoing research. Addressing these challenges is crucial for ensuring the reliability and effectiveness of AI systems in real-world disaster scenarios. The future of AI in disaster management lies in integrating with emerging technologies like the Internet of Things (IoT) to achieve more comprehensive and real-time monitoring of environmental conditions and infrastructure. Leveraging advanced data analytics and computational power can further improve the accuracy and timeliness of AI predictions and disaster simulations. Global collaboration and knowledge sharing are essential for developing standardized methods and shared technologies, along with establishing global databases of disaster-related data. Public–private partnerships can drive innovation and investment in AI for disaster management by supporting collaborative projects and research initiatives. Hence, the application of AI in natural disaster management is rapidly evolving, offering significant potential for improving prediction, monitoring, and response efforts. While challenges remain, ongoing research and development, coupled with strategic collaborations and investments, will pave the way for more effective, efficient, and equitable disaster management strategies in the future.

References

1. Lei, S., Wang, C., & Hou, Y. (2022). *Power grid resilience against natural disasters: Preparedness, response, and recovery*. Wiley.
2. Calvello, M. (2017). Early warning strategies to cope with landslide risk. *Rivista Italiana di Geotecnica, 2*(17), 63–91.
3. Stephens, K. K., Carlson, N. H., & Xu, Y. (2023). Disaster rescue communication using mobile devices, social media, and artificial intelligence. *International handbook of disaster research* (pp. 629–645). Springer.
4. Kikon, A., & Deka, P. C. (2022). Artificial intelligence application in drought assessment, monitoring and forecasting: A review. *Stochastic Environmental Research and Risk Assessment, 36*(5), 1197–1214.
5. Chauhan, B. V., Jaiswar, A., Bedi, A., Verma, S., Shrivastaw, V. K., & Vedrtnam, A. (2021). Applications of artificial intelligence and molecular immune pathogenesis, ongoing diagnosis and treatments for COVID-19. In *Artificial Intelligence for COVID-19* (pp. 521–549).
6. Díaz-Rodríguez, N., Del Ser, J., Coeckelbergh, M., de Prado, M. L., Herrera-Viedma, E., & Herrera, F. (2023). Connecting the dots in trustworthy Artificial Intelligence: From AI principles, ethics, and key requirements to responsible AI systems and regulation. *Information Fusion, 99*, 101896.
7. Albahri, A., Khaleel, Y. L., Habeeb, M. A., Ismael, R. D., Hameed, Q. A., Deveci, M., ... Alzubaidi, L. (2024). A systematic review of trustworthy artificial intelligence applications in natural disasters. *Computers and Electrical Engineering, 118*, 109409.
8. Sun, W., Bocchini, P., & Davison, B. D. (2020). Applications of artificial intelligence for disaster management. *Natural Hazards, 103*(3), 2631–2689.
9. Lorenz, P. (2020). *AI governance through political fora and standards developing organizations—Mapping the actors relevant to AI governance*. Mapping AI Governance Fora.
10. Ghaffarian, S., Taghikhah, F. R., & Maier, H. R. (2023). Explainable artificial intelligence in disaster risk management: Achievements and prospective futures. *International Journal of Disaster Risk Reduction, 98*, 104123.
11. Aydin, H. E., & Iban, M. C. (2023). Predicting and analyzing flood susceptibility using boosting-based ensemble machine learning algorithms with SHapley Additive exPlanations. *Natural Hazards, 116*(3), 2957–2991.
12. Mustafa, A. M., Agha, R., Ghazalat, L., & Sha'ban, T. (2024). Natural disasters detection using explainable deep learning. *Intelligent Systems with Applications, 23*, 200430.
13. Kumar, D., Bassill, N. P., & Ghosh, S. (2024). Analyzing recent trends in deep-learning approaches: A review on urban environmental hazards and disaster studies for monitoring, management, and mitigation toward sustainability. *International Journal on Smart Sensing and Intelligent Systems, 17*(1).
14. Tan, L., Guo, J., Mohanarajah, S., & Zhou, K. (2021). Can we detect trends in natural disaster management with artificial intelligence? A review of modeling practices. *Natural Hazards, 107*, 2389–2417.
15. Yavas, C. E., Chen, L., Kadlec, C., & Ji, Y. (2024). *Predictive modeling of earthquakes in Los Angeles with machine learning and neural networks*. IEEE Access.

16. Rouet-Leduc, B., Hulbert, C., Lubbers, N., Barros, K., Humphreys, C. J., & Johnson, P. A. (2017). Machine learning predicts laboratory earthquakes. *Geophysical Research Letters, 44*(18), 9276–9282.
17. Xu, C., & Xu, X. W. (2012). Spatial prediction models for seismic landslides based on support vector machine and varied kernel functions: A case study of the 14 April 2010 Yushu earthquake in China. *Chinese Journal of Geophysics, 55*(6), 666–679.
18. Asim, K., Martínez-Álvarez, F., Basit, A., & Iqbal, T. (2017). Earthquake magnitude prediction in Hindukush region using machine learning techniques. *Natural Hazards, 85*, 471–486.
19. Panakkat, A., & Adeli, H. (2007). Neural network models for earthquake magnitude prediction using multiple seismicity indicators. *International Journal of Neural Systems, 17*(01), 13–33.
20. DeVries, P. M., Viégas, F., Wattenberg, M., & Meade, B. J. (2018). Deep learning of aftershock patterns following large earthquakes. *Nature, 560*(7720), 632–634.
21. Kong, Q., Trugman, D. T., Ross, Z. E., Bianco, M. J., Meade, B. J., & Gerstoft, P. (2019). Machine learning in seismology: Turning data into insights. *Seismological Research Letters, 90*(1), 3–14.
22. Chowdhury, M., Rahaman, M. A., Islam, T. S., Sultana, R., & Rahman, A. (2023). *Prediction of earthquakes: A step towards predicting the unpredictable*. Brac University.
23. Saini, K., Kalra, S., & Sood, S. K. (2022). An integrated framework for smart earthquake prediction: IoT, fog, and cloud computing. *Journal of Grid Computing, 20*(2), 17.
24. Lv, H., Zeng, X., Bao, F., Xie, J., Lin, R., Song, Z., & Zhang, G. (2022). ADE-Net: A deep neural network for DAS earthquake detection trained with a limited number of positive samples. *IEEE Transactions on Geoscience and Remote Sensing, 60*, 1–11.
25. Jiao, Z., & Shan, X. (2024). A Bayesian approach for forecasting the probability of large earthquakes using thermal anomalies from satellite observations. *Remote Sensing, 16*(9), 1542.
26. Jia, J., & Ye, W. (2023). Deep learning for earthquake disaster assessment: Objects, data, models, stages, challenges, and opportunities. *Remote Sensing, 15*(16), 4098.
27. Plevris, V. (2024). AI-driven innovations in earthquake risk mitigation: A future-focused perspective. *Geosciences, 14*(9), 244.
28. Lakshmi, D., Tiwari, R. S., Dhanaraj, R. K., Kadry, S. (2024). *Explainable AI (XAI) for sustainable development: Trends and applications*. CRC Press.
29. Collini, E., Palesi, L. I., Nesi, P., Pantaleo, G., Nocentini, N., & Rosi, A. (2022). Predicting and understanding landslide events with explainable AI. *IEEE Access, 10*, 31175–31189.
30. Matin, S. S., & Pradhan, B. (2021). Earthquake-induced building-damage mapping using Explainable AI (XAI). *Sensors, 21*(13), 4489.
31. Ross, Z. E., Meier, M. A., & Hauksson, E. (2018). P wave arrival picking and first-motion polarity determination with deep learning. *Journal of Geophysical Research: Solid Earth, 123*(6), 5120–5129.
32. Mousavi, S. M., & Beroza, G. C. (2020). A machine-learning approach for earthquake magnitude estimation. *Geophysical Research Letters, 47*(1), e2019GL085976.
33. Perol, T., Gharbi, M., & Denolle, M. (2018). Convolutional neural network for earthquake detection and location. *Science Advances, 4*(2), e1700578.
34. Linville, L., Pankow, K., & Draelos, T. (2019). Deep learning models augment analyst decisions for event discrimination. *Geophysical Research Letters, 46*(7), 3643–3651.
35. Mignan, A., & Broccardo, M. (2019). *One neuron is more informative than a deep neural network for aftershock pattern forecasting*. arXiv preprint arXiv:1904.01983
36. Liu, B., Wen, H., Di, M., Huang, J., Liao, M., Yu, J., & Xiang, Y. (2024). Mapping and interpretability of aftershock hazards using hybrid machine learning algorithms. *Journal of Rock Mechanics and Geotechnical Engineering*.
37. Shen, D., Zhang, Q., Xu, T., Zhu, H., Zhao, W., Yin, Z., … Xiong, H. (2019). *Machine learning-enhanced realistic framework for real-time seismic monitoring—The winning solution of the 2017 international aftershock detection contest* (pp. 1–19). arXiv preprint arXiv:1911.09275
38. Zhang, X., Zhang, M., & Tian, X. (2021). Real-time earthquake early warning with deep learning: Application to the 2016 M 6.0 Central Apennines, Italy earthquake. *Geophysical Research Letters, 48*(5), 2020GL089394.

39. Mousavi, S. M., Ellsworth, W. L., Zhu, W., Chuang, L. Y., & Beroza, G. C. (2020). Earthquake transformer—An attentive deep-learning model for simultaneous earthquake detection and phase picking. *Nature Communications, 11*(1), 3952.
40. Jia, Z., Zhan, Z., & Kanamori, H. (2022). The 2021 South Sandwich Island Mw 8.2 earthquake: A slow event sandwiched between regular ruptures. *Geophysical Research Letters, 49*(3), e2021GL097104.
41. Pwavodi, J., Ibrahim, A. U., Pwavodi, P. C., Al-Turjman, F., & Mohand-Said, A. (2024). The role of artificial intelligence and IoT in prediction of earthquakes. *Artificial Intelligence in Geosciences*, 100075.
42. Fayaz, J., & Galasso, C. (2024). Interpretability and spatial efficacy of a deep-learning-based on-site early warning framework using explainable artificial intelligence and geographically weighted random forests. *Geoscience Frontiers, 15*(5), 101839.
43. Karpatne, A., Atluri, G., Faghmous, J. H., Steinbach, M., Banerjee, A., Ganguly, A., … Kumar, V. (2017). Theory-guided data science: A new paradigm for scientific discovery from data. *IEEE Transactions on Knowledge and Data Engineering, 29*(10), 2318–2331.
44. Barnes, C. R., Best, M. M., Johnson, F. R., & Pirenne, B. (2010). Large subsea observatory for earth-ocean science: Challenges of multidisciplinary integration across hardware, software, and people networks. In *Oceans 2010 MTS/IEEE Seattle*. IEEE.
45. Huang, L., Li, J., Hao, H., & Li, X. (2018). Micro-seismic event detection and location in underground mines by using Convolutional Neural Networks (CNN) and deep learning. *Tunnelling and Underground Space Technology, 81*, 265–276.
46. Pugliese Viloria, A. d. J., Folini, A., Carrion, D., & Brovelli, M. A. (2024). Hazard susceptibility mapping with machine and deep learning: A literature review. *Remote Sensing, 16*(18), 3374.
47. Mosavi, A., Ozturk, P., & Chau, K.-W. (2018). Flood prediction using machine learning models: Literature review. *Water, 10*(11), 1536.
48. Tehrany, M. S., Pradhan, B., & Jebur, M. N. (2014). Flood susceptibility mapping using a novel ensemble weights-of-evidence and support vector machine models in GIS. *Journal of Hydrology, 512*, 332–343.
49. Muñoz, P., Orellana-Alvear, J., Willems, P., & Célleri, R. (2018). Flash-flood forecasting in an Andean mountain catchment—Development of a step-wise methodology based on the random forest algorithm. *Water, 10*(11), 1519.
50. Kratzert, F., Klotz, D., Brenner, C., Schulz, K., & Herrnegger, M. (2018). Rainfall–runoff modelling using long short-term memory (LSTM) networks. *Hydrology and Earth System Sciences, 22*(11), 6005–6022.
51. Nevo, S., Morin, E., Gerzi Rosenthal, A., Metzger, A., Barshai, C., Weitzner, D., … Dror, G. (2022). Flood forecasting with machine learning models in an operational framework. *Hydrology and Earth System Sciences, 26*(15), 4013–4032.
52. Tien Bui, D., Tuan, T. A., Klempe, H., Pradhan, B., & Revhaug, I. (2016). Spatial prediction models for shallow landslide hazards: A comparative assessment of the efficacy of support vector machines, artificial neural networks, kernel logistic regression, and logistic model tree. *Landslides, 13*, 361–378.
53. Peng, B., Meng, Z., Huang, Q., & Wang, C. (2019). Patch similarity convolutional neural network for urban flood extent mapping using bi-temporal satellite multispectral imagery. *Remote Sensing, 11*(21), 2492.
54. Huang, X., Wang, C., & Li, Z. (2018). A near real-time flood-mapping approach by integrating social media and post-event satellite imagery. *Annals of GIS, 24*(2), 113–123.
55. Nemni, E., Bullock, J., Belabbes, S., & Bromley, L. (2020). Fully convolutional neural network for rapid flood segmentation in synthetic aperture radar imagery. *Remote Sensing, 12*(16), 2532.
56. Sit, M., Demiray, B. Z., Xiang, Z., Ewing, G. J., Sermet, Y., & Demir, I. (2020). A comprehensive review of deep learning applications in hydrology and water resources. *Water Science and Technology, 82*(12), 2635–2670.

57. Feng, Q., Liu, J., & Gong, J. (2015). Urban flood mapping based on unmanned aerial vehicle remote sensing and random forest classifier—A case of Yuyao, China. *Water, 7*(4), 1437–1455.
58. Saravi, S., Kalawsky, R., Joannou, D., Rivas Casado, M., Fu, G., & Meng, F. (2019). Use of artificial intelligence to improve resilience and preparedness against adverse flood events. *Water, 11*(5), 973.
59. Perera, D., Agnihotri, J., Seidou, O., & Djalante, R. (2020). Identifying societal challenges in flood early warning systems. *International Journal of Disaster Risk Reduction, 51*, 101794.
60. Shen, X., Wang, D., Mao, K., Anagnostou, E., & Hong, Y. (2019). Inundation extent mapping by synthetic aperture radar: A review. *Remote Sensing, 11*(7), 879.
61. Wang, R.-Q., Mao, H., Wang, Y., Rae, C., & Shaw, W. (2018). Hyper-resolution monitoring of urban flooding with social media and crowdsourcing data. *Computers & Geosciences, 111*, 139–147.
62. Khosravi, K., Shahabi, H., Pham, B. T., Adamowski, J., Shirzadi, A., Pradhan, B., … Ho, H. L. (2019). A comparative assessment of flood susceptibility modeling using multi-criteria decision-making analysis and machine learning methods. *Journal of Hydrology, 573*, 311–323.
63. Tehrany, M. S., Kumar, L., & Drielsma, M. J. (2017). Review of native vegetation condition assessment concepts, methods and future trends. *Journal for Nature Conservation, 40*, 12–23.
64. Khosravi, K., Pham, B. T., Chapi, K., Shirzadi, A., Shahabi, H., Revhaug, I., … Bui, D. T. (2018). A comparative assessment of decision trees algorithms for flash flood susceptibility modeling at Haraz watershed, northern Iran. *Science of the Total Environment, 627*, 744–755.
65. Pham, B. T., Luu, C., Van Phong, T., Nguyen, H. D., Van Le, H., Tran, T. Q., … Prakash, I. (2021). Flood risk assessment using hybrid artificial intelligence models integrated with multi-criteria decision analysis in Quang Nam Province, Vietnam. *Journal of Hydrology, 592*, 125815.
66. Wagenaar, D., Lüdtke, S., Schröter, K., Bouwer, L. M., & Kreibich, H. (2018). Regional and temporal transferability of multivariable flood damage models. *Water Resources Research, 54*(5), 3688–3703.
67. Zischg, A. P., Mosimann, M., Bernet, D. B., & Röthlisberger, V. (2018). Validation of 2D flood models with insurance claims. *Journal of Hydrology, 557*, 350–361.
68. Sit, M. A., Koylu, C., & Demir, I. (2020). Identifying disaster-related tweets and their semantic, spatial and temporal context using deep learning, natural language processing and spatial analysis: A case study of Hurricane Irma. *Social Sensing and Big Data Computing for Disaster Management* (pp. 8–32). Routledge.
69. Amit, S. N. K. B., & Aoki, Y. (2017). Disaster detection from aerial imagery with convolutional neural network. In *2017 international electronics symposium on knowledge creation and intelligent computing (IES-KCIC)*. IEEE.
70. Sublime, J., & Kalinicheva, E. (2019). Automatic post-disaster damage mapping using deep-learning techniques for change detection: Case study of the Tohoku tsunami. *Remote Sensing, 11*(9), 1123.
71. Kharazi, B. A., & Behzadan, A. H. (2021). Flood depth mapping in street photos with image processing and deep neural networks. *Computers, Environment and Urban Systems, 88*, 101628.
72. Nojavan, F., Qian, S. S., & Stow, C. A. (2017). Comparative analysis of discretization methods in Bayesian networks. *Environmental Modelling & Software, 87*, 64–71.
73. Nguyen, D. T., Ofli, F., Imran, M., & Mitra, P. (2017). Damage assessment from social media imagery data during disasters. In *Proceedings of the 2017 IEEE/ACM International Conference on Advances in Social Networks Analysis and Mining.*
74. Boussioux, L., Zeng, C., Guénais, T., & Bertsimas, D. (2022). Hurricane forecasting: A novel multimodal machine learning framework. *Weather and Forecasting, 37*(6), 817–831.
75. Chen, R., Wang, X., Zhang, W., Zhu, X., Li, A., & Yang, C. (2019). A hybrid CNN-LSTM model for typhoon formation forecasting. *GeoInformatica, 23*, 375–396.
76. Giffard-Roisin, S., Yang, M., Charpiat, G., Kumler Bonfanti, C., Kégl, B., & Monteleoni, C. (2020). Tropical cyclone track forecasting using fused deep learning from aligned reanalysis data. *Frontiers in Big Data, 3*, 1.

77. Mudigonda, M., Kim, S., Mahesh, A., Kahou, S., Kashinath, K., Williams, D., … Prabhat, M. (2017). Segmenting and tracking extreme climate events using neural networks. In *Deep Learning for Physical Sciences (DLPS) Workshop, Held with NIPS Conference.*
78. Kumler-Bonfanti, C., Stewart, J., Hall, D., & Govett, M. (2020). Tropical and extratropical cyclone detection using deep learning. *Journal of Applied Meteorology and Climatology, 59*(12), 1971–1985.
79. Lee, J., Im, J., Cha, D.-H., Park, H., & Sim, S. (2019). Tropical cyclone intensity estimation using multi-dimensional convolutional neural networks from geostationary satellite data. *Remote Sensing, 12*(1), 108.
80. Wimmers, A., Velden, C., & Cossuth, J. H. (2019). Using deep learning to estimate tropical cyclone intensity from satellite passive microwave imagery. *Monthly Weather Review, 147*(6), 2261–2282.
81. Li, Z., Y. Wen, M. Schreier, A. Behrangi, Y. Hong, & B. Lambrigtsen (2021). Advancing satellite precipitation retrievals with data driven approaches: Is black box model explainable? *Earth and Space Science, 8*(2), e2020EA001423.
82. Leroux, M.-D., Wood, K., Elsberry, R. L., Cayanan, E. O., Hendricks, E., Kucas, M., … Yu, Z. (2018). Recent advances in research and forecasting of tropical cyclone track, intensity, & structure at landfall. *Tropical Cyclone Research and Review, 7*(2), 85–105.
83. Moradi Kordmahalleh, M., Gorji Sefidmazgi, M., & Homaifar, A. (2016). A sparse recurrent neural network for trajectory prediction of Atlantic Hurricanes. In *Proceedings of the Genetic and Evolutionary Computation Conference.*
84. Chattopadhyay, A., Hassanzadeh, P., & Pasha, S. (2020). Predicting clustered weather patterns: A test case for applications of convolutional neural networks to spatio-temporal climate data. *Scientific Reports, 10*(1), 1317.
85. Soni, J., Prabakar, N., & Upadhyay, H. *A transfer learning approach for hurricane damage assessment using satellite imagery.*
86. Li, J., Ozbay, K., Bartin, B., Iyer, S., & Carnegie, J. A. (2013). Empirical evacuation response curve during hurricane irene in cape may county, new jersey. *Transportation Research Record, 2376*(1), 1–10.
87. Tadesse, M., Wahl, T., & Cid, A. (2020). Data-driven modeling of global storm surges. *Frontiers in Marine Science, 7*, 512653.
88. Wang, J., Kuffer, M., Roy, D., & Pfeffer, K. (2019). Deprivation pockets through the lens of convolutional neural networks. *Remote Sensing of Environment, 234*, 111448.
89. Doshi, J., Basu, S., & Pang, G. (2018). *From satellite imagery to disaster insights.* arXiv preprint arXiv:1812.07033
90. Rahnemoonfar, M., Murphy, R., Miquel, M. V., Dobbs, D., & Adams, A. (2018). Flooded area detection from UAV images based on densely connected recurrent neural networks. In *IGARSS 2018–2018 IEEE International Geoscience and Remote Sensing Symposium.* IEEE.
91. Hao, H., & Wang, Y. (2020). Leveraging multimodal social media data for rapid disaster damage assessment. *International Journal of Disaster Risk Reduction, 51*, 101760.
92. Ren, C. X., Peltier, A., Ferrazzini, V., Rouet-Leduc, B., Johnson, P. A., & Brenguier, F. (2020). Machine learning reveals the seismic signature of eruptive behavior at Piton de la Fournaise volcano. *Geophysical Research Letters, 47*(3), e2019GL085523.
93. Valade, S., Ley, A., Massimetti, F., D'Hondt, O., Laiolo, M., Coppola, D., … Walter, T. R. (2019). Towards global volcano monitoring using multisensor sentinel missions and artificial intelligence: The MOUNTS monitoring system. *Remote Sensing, 11*(13), 1528.
94. Titos, M., Bueno, A., Garcia, L., & Benitez, C. (2018). A deep neural networks approach to automatic recognition systems for volcano-seismic events. *IEEE Journal of Selected Topics in Applied Earth Observations and Remote Sensing, 11*(5), 1533–1544.
95. Parra, R. (2018). Numerical system for forecasting volcanic ash dispersion in Ecuador. In *2018 IEEE Third Ecuador Technical Chapters Meeting (ETCM).* IEEE.
96. Cannavò, F., Cannata, A., Cassisi, C., Di Grazia, G., Montalto, P., Prestifilippo, M., … Gambino, S. (2017). A multivariate probabilistic graphical model for real-time volcano monitoring on Mount Etna. *Journal of Geophysical Research: Solid Earth, 122*(5), 3480–3496.

97. Anantrasirichai, N., Biggs, J., Albino, F., Hill, P., & Bull, D. (2018). Application of machine learning to classification of volcanic deformation in routinely generated InSAR data. *Journal of Geophysical Research: Solid Earth, 123*(8), 6592–6606.
98. Gaddes, M., Hooper, A., Bagnardi, M., Inman, H., & Albino, F. (2018). Blind signal separation methods for InSAR: The potential to automatically detect and monitor signals of volcanic deformation. *Journal of Geophysical Research: Solid Earth, 123*(11), 10226–10251.
99. Aldeghi, A., Carn, S., Escobar-Wolf, R., & Groppelli, G. (2019). Volcano monitoring from space using high-cadence planet CubeSat images applied to Fuego volcano, Guatemala. *Remote Sensing, 11*(18), 2151.
100. Reath, K., Pritchard, M., Poland, M., Delgado, F., Carn, S., Coppola, D., … Henderson, S. (2019). Thermal, deformation, and degassing remote sensing time series (CE 2000–2017) at the 47 most active volcanoes in Latin America: Implications for volcanic systems. *Journal of Geophysical Research: Solid Earth, 124*(1), 195–218.
101. Hébert, H., Occhipinti, G., Schindelé, F., Gailler, A., Pinel-Puysségur, B., Gupta, H., … Meilianda, E. (2020). Contributions of space missions to better tsunami science: observations, models and warnings. *Surveys in Geophysics, 41*(6), 1535–1581.
102. Bueno, A., Benítez, C., De Angelis, S., Moreno, A. D., & Ibáñez, J. M. (2019). Volcano-seismic transfer learning and uncertainty quantification with Bayesian neural networks. *IEEE Transactions on Geoscience and Remote Sensing, 58*(2), 892–902.
103. Romeo, F., Mereu, L., Scollo, S., Papa, M., Corradini, S., Merucci, L., & Marzano, F. S. (2023). Volcanic cloud detection and retrieval using satellite multisensor observations. *Remote Sensing, 15*(4), 888.
104. Ghaffarian, S., Kerle, N., & Filatova, T. (2018). Remote sensing-based proxies for urban disaster risk management and resilience: A review. *Remote Sensing, 10*(11), 1760.
105. Jumadi, Heppenstall, A. J., Malleson, N. S., Carver, S. J., Quincey, D. J., & Manville, V. R. (2018). Modelling individual evacuation decisions during natural disasters: a case study of volcanic crisis in Merapi, Indonesia. *Geosciences, 8*(6), 196.
106. Deligne, N. I., Fitzgerald, R. H., Blake, D. M., Davies, A. J., Hayes, J. L., Stewart, C., … Kennedy, B. M. (2017). Investigating the consequences of urban volcanism using a scenario approach I: Development and application of a hypothetical eruption in the Auckland Volcanic Field, New Zealand. *Journal of Volcanology and Geothermal Research, 336*, 192–208.
107. Pallister, J., Papale, P., Eichelberger, J., Newhall, C., Mandeville, C., Nakada, S., … Ewert, J. (2019). Volcano observatory best practices (VOBP) workshops-a summary of findings and best-practice recommendations. *Journal of Applied Volcanology, 8*(1), 1–33.
108. Que-Salinas, S. U., Torres-Orozco, R., Sieron, K., Juárez-Cerrillo, S. F., & Córdoba-Montiel, F. *Artificial neural networks-based 'first-response' detection of volcano-tectonic seismic events in poorly monitored active volcanoes: The case of Pico De Orizaba (Citlaltépetl)*, Mexico.
109. Cardwell, R., McDonald, G., Wotherspoon, L., & Lindsay, J. (2021). Simulation of post volcanic eruption land use and economic recovery pathways over a period of 20 years in the Auckland region of New Zealand. *Journal of Volcanology and Geothermal Research, 415*, 107253.
110. Bonadonna, C., Frischknecht, C., Menoni, S., Romerio, F., Gregg, C. E., Rosi, M., … Guobadia, D. (2021). Integrating hazard, exposure, vulnerability and resilience for risk and emergency management in a volcanic context: the ADVISE model. *Journal of Applied Volcanology, 10*, 1–34.
111. Ferrari, E., Massa, M., Lovati, S., Di Michele, F., & Rizzo, A. L. (2024). Multiparametric stations for real-time monitoring and long-term assessment of natural hazards. *Frontiers in Earth Science, 12*, 1412900.
112. Bail, R. d. F., Kovaleski, J. L., da Silva, V. L., Pagani, R. N., & Chiroli, D. M. d. G. (2021). Internet of things in disaster management: Technologies and uses. *Environmental Hazards, 20*(5), 493–513.
113. Ghapar, A. A., Yussof, S., & Bakar, A. A. (2018). Internet of Things (IoT) architecture for flood data management. *International Journal of Future Generation Communication and Networking, 11*(1), 55–62.

114. Prakash, C., Barthwal, A., & Acharya, D. (2022). FLOODWALL: A real-time flash flood monitoring and forecasting system using IoT. *IEEE Sensors Journal, 23*(1), 787–799.
115. Goyal, H. R., Ghanshala, K. K., & Sharma, S. (2021). Post flood management system based on smart IoT devices using AI approach. *Materials Today: Proceedings, 46*, 10411–10417.
116. Yu, M., Yang, C., & Li, Y. (2018). Big data in natural disaster management: A review. *Geosciences, 8*(5), 165.
117. Qadir, J., Ali, A., ur Rasool, R., Zwitter, A., Sathiaseelan, A., & Crowcroft, J. (2016). Crisis analytics: big data-driven crisis response. *Journal of International Humanitarian Action, 1*, 1–21.
118. Khan, S. M., Shafi, I., Butt, W. H., Diez, I. d. l. T., Flores, M. A. L., Galán, J. C., & Ashraf, I. (2023). A systematic review of disaster management systems: approaches, challenges, and future directions. *Land, 12*(8), 1514.
119. Manfré, L. A., Hirata, E., Silva, J. B., Shinohara, E. J., Giannotti, M. A., Larocca, A. P. C., & Quintanilha, J. A. (2012). An analysis of geospatial technologies for risk and natural disaster management. *ISPRS International Journal of Geo-Information, 1*(2), 166–185.
120. Ray-Bennett, N., Mendez, D., Alam, E., & Morgner, C. (2020). Inter-agency collaboration for natural hazard management in developed countries. In *Oxford research encyclopedia of natural hazard science*. Oxford University Press.
121. Linardos, V., Drakaki, M., Tzionas, P., & Karnavas, Y. L. (2022). Machine learning in disaster management: recent developments in methods and applications. *Machine Learning and Knowledge Extraction, 4*(2).
122. Gupta, S., Modgil, S., Kumar, A., Sivarajah, U., & Irani, Z. (2022). Artificial intelligence and cloud-based collaborative platforms for managing disaster, extreme weather and emergency operations. *International Journal of Production Economics, 254*, 108642.
123. Mariani, M. M., Machado, I., Magrelli, V., & Dwivedi, Y. K. (2023). Artificial intelligence in innovation research: A systematic review, conceptual framework, and future research directions. *Technovation, 122*, 102623.

Chapter 7
The Dark Side of AI: Negative Environmental Impacts of AI Advancements

Tanmoy Banerjea, M. K. Shukla, Balendra V. S. Chauhan, and Sneha Verma

Abstract Artificial intelligence, with its powerful ability to process vast data sets and generate rapid outcomes, can rightfully be linked to complex and statistical simulations at its core. Even though at the time of its inception, it started off as a humble enabler in man's life, today in the twenty-first century there is no doubt about the tremendous expanse of its meaningful allure that has successfully captivated its unsolicited place in almost anything that we can think of all in a short while since its evolution. AI's ability to analyse patterns and leverage vast sets of knowledge has set the stage for upcoming improvements in energy production, conversion, and delivery, which has without doubt, made modern ways of living extremely convenient. However, the rapid advancements of AI carry hidden costs, contributing to environmental degradation in unforeseen ways. This chapter consists of four different aspects of the use and effect of AI such as—electronic waste generation; excessive resource consumption; self-driving cars; and the added pressure on rare earth elements –all of which are keystone components of environmental degradation. Scientists too have come to recognize its negative impacts at an early stage and here we have discussed some conventional practices that such systems generally resort to, along with their estimated outcomes to give the readers a glimpse of how things happen.

Keywords Artificial Intelligence · e-Waste · Self-driving car · Rare earth elements

B. V. S. Chauhan (✉)
Centre for Earth Observation Science, University of Brighton, Brighton, UK
e-mail: balendrachauhan8801@gmail.com

T. Banerjea · M. K. Shukla
CSIR, Indian Institute of Petroleum (CSIR-IIP), Dehradun, India
e-mail: banerjeatanmoy@gmail.com

M. K. Shukla
e-mail: mshukla@iip.res.in

B. V. S. Chauhan · S. Verma
School of Engineering, Newcastle University, Newcastle Upon Tyne, United Kingdom
e-mail: Sneha.verma@newcastle.ac.uk

A. Vedrtnam et al., *Prospects of Artificial Intelligence in the Environment*,
https://doi.org/10.1007/978-981-96-6863-2_7

1 Introduction

Technological changes are fundamental to scientific and economic breakthroughs [1] and they have the potential to dramatically influence global efforts toward sustainability [2, 3]. One could therefore comprehend the importance of machine learning (ML) models in modern-day technology from the surging popularity of commercial AI products based on their generative and multi-purpose utility. Such systems have shown promise in solving complex data patterns or formats [4] and newer ways of data processing frameworks in order to get meaningful insights, which are and have been unprecedented [5].

Machine learning models and by extension, artificial intelligence, are currently building upon concepts of improved coordination and automation to improve application features, designs and testing. Their ongoing improvements are noticeable in an almost limitless market, especially in the United States of America [6] and the developed West.

In animal farming, mechanistic models have the potential to solve complex problems such as: identifying functional limiting factors, determining optimal nutrient composition of animal feed, evaluating animal management to evaluate performance [7], examining strategies to reduce nutrient excretion into the environment [8], etc. All these make use of sensors—devices that measure or detect a biological, chemical, physical or mechanical property or a combination thereof, to record and collect data for interpretation by a human or a machine [9].

Even in the field of healthcare, machine learning can be used gainfully to identify the optimal trial sample, collect more data points, assess ongoing data from trial participants, and eliminate data-based errors [10]; predict diseases, such as diabetes—a prevalent and chronic condition affecting millions of people worldwide—whose early detection and trend prediction can significantly impact patient outcomes by enabling timely interventions, personalized treatment plans, and improved disease management [12]; detecting early indicators of an epidemic or pandemic through algorithms that examine satellite data, news and social media reports, and even video sources to determine the volatility of the sickness. Using ML for healthcare can therefore open up a world of possibilities in this field [10].

However, the concept of "generality" has its fair share of financial and environmental costs, given the amount of energy that these systems require and the amount of carbon they emit. In other words, training and running complex algorithms and models for deep learning and cloud computing consume vast volumes of energy [11], a fact that is best represented by the equation shown below:

$$\text{Computing Time} \propto [\text{Power Input}]^{(-1)}$$

LHS—unit in milliseconds/microseconds

RHS—unit in watts/kilowatts.

Automated decision making and predictive analyses through artificial intelligence in combination with rapid progress in technologies such as sensor technology and robotics are likely to change the way individuals, communities, governments and other entities perceive and respond to climate and ecological change [13]. Studies conducted in the latter half of the covid-19 era have revealed some key approximate findings, with the first being a minimum requirement of 1300 MW of energy for training GPT-3,thatmay be equivalent to powering 130 homes in United States of America [14]; the second one being the more expensive nature of multi-purpose, generative architectures as compared to task specific systems for a variety of tasks (even while controlling for the number of model parameters) [15]; and power usage for text prompt operations being of the order of a few milli-watt hours per task as compared to a few watt hours for a set of a thousand text generations [16] being yet another. Not to forget the exponentially intensive power usage for image generation on AI platforms, which may be equivalent to charging a smart phone for one whole cycle [17], as compared to the moderate nature of text prompt operations, that has unfortunately gotten data centres across the globe to account for 1–2% of global electricity (approximated to 400 terawatt hours of power per year, putting it at the same level as the standalone energy consumption needs of some European countries [18]), the future forecasts project this demand to go up to 800 terawatt hours per year, by 2026 [18]. Needless to say, the figures speak volumes for themselves by painting an alarming picture of mankind's prospective consumption patterns.

2 Electronic Waste

By definition, electronic wastes encompass many forms of electronic products from personal gadgets to home appliances that have lost value and are considered to be useless [19]. They are generated in enormous amounts worldwide and the lack of appropriate ways of dealing with them, especially in developing countries, have made this a global environmental issue. Africa as a whole is undoubtedly the fastest growing economy than any other continent and has seen a rapid increase in imports of electronic and electrical goods to cater to the implementation of information and communication (ICT) technologies [20].

It was estimated that 41.8 million tonnes of e-waste were produced globally in 2014 [21], but as per an assessment [22] the average lifespan of a new computer decreased from 4.5 years in 1992 to an estimated 2 years in 2005, and is further decreasing. This is bound to result in the disposal of more computers, or at best, having them exported to developing countries. Apart from that, the toxic substances produced by e-wastes which include lead, cadmium, mercury, and brominated flame retardants (BFRs) such as poly-brominated biphenyl ethers (PBDEs) and hexa-bromo-cyclodo-decanes, are mostly disposed of by open burning in most developing countries, thus putting scores of communities and environs at risk of exposure to carcinogenic releases, such as those of dioxins and toxic metals, into the environment. They have significant abilities of contaminating any surrounding area in which they are placed, and threatening human health if their ‘end-of-life’ management is not done meticulously. Current trends show the total quantity of waste electrical and electronic equipment (WEEEs) increasing at a rate of 2.6 million tonnes annually, with 78.4 million tonnes being the 2030 global estimation target of electronic waste production. This, coupled with abysmally low recycling rates of 17.4%, is highly concerning [19]. With the World Economic Forum (WEF) also projecting a generation of 120 million metric tonnes of e-waste by the year 2050, it is high time we set the alarm bells ringing.

According to the definitions in the Directive 2002/96/EC of the European Parliament and of the Council (January 2003) on Waste Electrical and Electronic Equipment, WEEE consists of the ten categories listed in Table 1.

Though considered a ticking time bomb, a process called e-waste recycling in which valuable materials are extracted after shredding the e-waste into tiny pieces that could be used once again in new electronic appliances, poses to be a potential solution to get started with. For that, one must keep track of some undeniable facts [23], such as: (1) the presence of 347 metric tons of non recycled e-waste on earth currently; (2) collection and proper recycling of only 17.4% of the total known e-waste presently; (3) China (10129 Kt), USA (6918 Kt) and India (3230 Kt) together being the highest producers of e-waste, with their respective recycling rates being at 16, 15 and 1% [23], (4) less than 80 countries across the world having a dedicated legislation for dealing with e-wastes within their geographical borders; and (5) 3 out of 27 countries in the Schengen area, namely Estonia, Norway and Iceland having the highest e-waste recycling rates. The same study also shows: a global e-waste generation of 44.4 Mt in 2014, 57.4 Mt in 2021, 59.4 Mt in 2022, 65.3 Mt in 2023, and an estimated 74.7 Mt by 2030; Asia, Americas and Europe producing 29.4, 13.1 and 12 Mt of e-waste by volume, respectively; Europe, Oceania and Americas producing 16.2, 16.1 and 13.3 kg of e-waste per capita respectively; categories of e-waste—small equipment (17.4

Table 1 WEEE categories according to the EU directive on WEEE (EU 2002a)

S. No.	Category	Label
1	Large household appliances	Large HH
2	Small household appliances	Small HH
3	IT and telecommunications equipment	ICT
4	Consumer equipment	CE
5	Lighting equipment	Lighting
6	Electrical and electronic tools (with the exception of large-scale stationary industrial tools)	E & E tools
7	Toys, leisure and sports equipment	Toys
8	Medical devices (with the exception of all implanted and infected products)	Medical equipment
9	Monitoring and control instruments	M & C
10	Automatic dispensers	Dispensers

Table 2 Average abundance (in μg/g) of REEs in the earth's crust as compared to chondritic abundances [58]

Elements	Taylor and MeLennan (1985)	Wedepohl (1995)	Lide (1997)	Wakita et al. (1971) (Chondritic Abundances)	Pourmand et al. (2012) (Chondritic Abundances)
La	16.0	30	39	0.34	0.2469
Ce	33.0	60	66.5	0.91	0.6321
Pr	3.9	6.7	9.2	0.121	0.0959
Nd	16.0	27	41.5	0.64	0.4
Sm	3.5	5.3	7.05	0.195	0.1556
Eu	1.1	1.3	2	0.073	0.0599
Gd	3.3	4	6.2	0.26	0.2093
Tb	0.6	0.65	1.2	0.047	0.0378
Dy	3.7	3.8	5.2	0.3	0.2577
Ho	0.8	0.8	1.3	0.078	0.0554
Er	2.2	2.1	3.5	0.2	0.1667
Tm	0.3	0.3	0.52	0.032	0.0261
Yb	2.2	2	3.2	0.22	0.1694
Lu	0.3	0.35	0.8	0.034	0.0256
Y	20.0	24	33	-	1.395
Sc	30.0	16	22	-	5.493
Total	136.9	194.3	242.17	-	9.5118

Note "Chondritic Abundances" is a term that represents universal abundance of such elements, as observed and retrieved from meteorites that fly in from outer space

Mt), large equipment (13.1 Mt), temperature exchange equipment (10.8 Mt), screens and monitors (6.7 Mt), small IT and telecoms equipment (4.7 Mt), and illuminous objects (0.9 Mt) in 2023.

Note Kt stands for kilo tons; Mt stands for million metric tons

We thus understand that e-waste management (EWM) is a global challenge with complexities beyond general waste management [24, 25], which when done inappropriately would give rise to issues such lack of energy efficiency, magnified carbon footprint, heterogeneity, poor technology management and all-round administration, that can impact supply chains [26]. Taking note of vast expanses of the cause and effect relationship between e-wastes and our environment, some immediate counter measures should include a few things such as—(1) a push for legally binding obligations at international summits and incentivizing voluntary adherence to the provisions of their joint statements e.g., conventions similar to "Polluter Pays"; (2) funding critically important research & development initiatives, asides the dissemination of awareness in waste recycling and it's management; (3) proliferation of the technical know-how of e-waste management in countries that are heavily reliant on virtual technologies and solar power; etc. As global problems have always had a history of evolving solutions based on international consensus, the ideal way forward should inevitably follow the same course of action, as has been so after the 1950s.

3 Resource Consumption

Some revelations made by researchers at OpenAI deserve a mention, if not a critical analysis. As per one such assessment [27] the amount of computing power needed to train cutting-edge AI models has doubled almost every 4 months since 2012, which may raise serious questions on the widely accepted Moore's Law, that states an approximately 2x (i.e., 100%) increment in the number of transistors on an integrated circuit (IC) every two years. The same study [27] also indicates that emissions from the ICT industry as a whole (primarily including infrastructure) are expected to touch 14% of the global emissions by year 2040. They give us a relatively true depiction of the periodic pattern of resource consumption that is in the making.

Of the several dependencies that define our potential of survival on Earth, there are two very important resources that control the economic and social potential of twenty-first century habitations, which also happen to be keystone components of development in the Atomic Age. They are: energy and water; and are among our primary concern.

3.1 Energy

Ever since its discovery, the possession of its reserves in different forms has been a vital parameter for recognizing nations as superpowers and regional powers, primarily because of the socio-economic fabric of the world in an era after the nineteenth century. And ever since, there has been no looking back as is evident from the space it has created for itself in our modern needs and advancements. A study conducted by researchers at the University of Massachusetts prior to 2019 gave the world an estimate of the amount of energy that is, on an average, used to train certain popular large artificial intelligence models [28] (Fig. 1).

The results, as shown in the above figure, report a production of nearly 626,000 lbs of carbon dioxide, which is similar to taking 300 round-trip flights between New York City and San Francisco. That is equivalent to nearly 5 times the lifetime emissions of an average car.

A study done by [29] has reviewed existing tools for estimating the energy consumption of computations, and explains the specific notions that are not always known by AI practitioners. The seven different tools that we study are: Green-Algorithms; Code Carbon (CC (M) for machine); Eco2AI (E2 (P) for process, E2 (M) for machine); Carbon Tracker; Experiment-Impact-Tracker (EIT); MLCO2 and Cumulator (CMLTRs). Using these tools, the energy consumed by CPU, GPU,

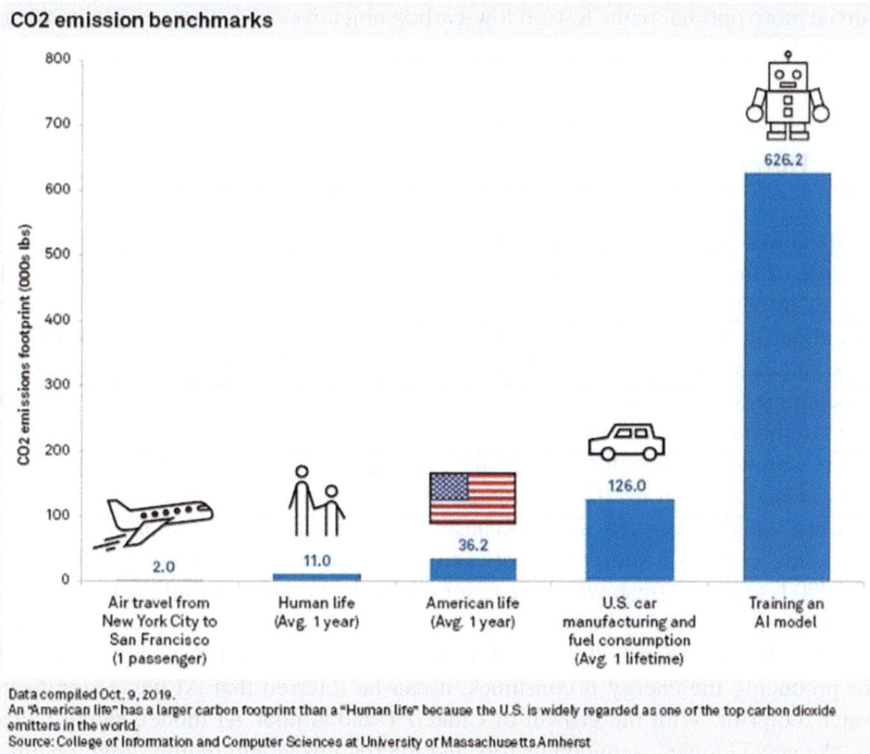

Fig. 1 Estimated emissions in different activities (of an average American)

memory and communications can be estimated. The most power consuming devices on a personal machine or a server are the GPUs (if present), CPUs, and memory. Conservative estimates state that data centres will consume 4.5% of global energy by 2030 (Patel, Nishball and Ontiveros 2024) [30].

Note Some tools had to be modified to be used, as they had bugs not yet fixed by the authors.

Considering the vast environmental consequences associated with the development and use of AI, of which the world is only starting to learn, the necessity of addressing AI alongside the concept of infrastructure points toward the phenomenon of carbon lock-in [31]. Carbon lock-in refers to society's constrained ability to reduce carbon emissions technologically, economically, politically, and socially. These constraints are due to the inherent inertia created by entrenched technological, institutional, and behavioural norms. The phenomenon of lock-in is most referenced in terms of carbon lock-in and the concern for greenhouse gas (GHG) emissions [34]. Carbon Lock-in refers to "the dynamic whereby prior decisions relating to GHG-emitting technologies, infrastructure, practices, and their supporting networks constrain future paths, making it more challenging, even impossible, to subsequently pursue more optimal paths toward low-carbon objectives". Coal power plants are an oft-cited example of a carbon lock-in [34, 35].

3.2 Water

Civilizations have, since time immemorial, been able to thrive on Earth solely because of the presence of water in abundant measures. That alone should emphasize its importance and the need for its conservation. But AI could certainly care less about it. Fresh (potable) water is an important contributor in the training of large language modules for deep learning, cloud computing and complex algorithms. Supercomputers and conventional computers are radiators which convert every unit of consumed energy into heat and therefore, water is used to keep the servers cool. The reason why it is preferred over air is because of its greater thermal capacity to transfer heat through itself as compared to the latter. At the same time, fresh water gets precedence over sea water because of the sensitivity of equipment involved in cooling systems towards minerals and salinity. Several studies have shown that 700,000 L of clean, fresh water are required for the purpose of training GPT-3 [32], which should preferably be cold.

Since AI uses water both for cooling the servers that power its computations and for producing the energy it consumes, it can be inferred that AI has a significant water footprint. With the growth of ChatGPT and similar AI models being hailed as "the new Google", semiconductors and microchips would require large volumes of water in the manufacturing stage itself. And this is only expected to grow in the future. But while a single Google search requires half a millilitre of water in energy, ChatGPT consumes 500 millilitres of water for every five to 50 prompts [33]. AI uses and pollutes water through related hardware production. Producing the AI hardware involves resource-intensive mining for rare materials such as silicon, germanium, gallium, boron and phosphorous, the extraction of which has significant impacts on the environment and contributes to water pollution. Data centres provide the physical infrastructure for training and running AI, and their energy consumption could double by 2026. Technology firms using water to run and cool these data centres potentially require water withdrawals of 4.2–6.6 billion cubic metres by 2027. By comparison, Google's data centres used over 21 billion litres of potable water in 2022, an increase of 20% on its 2021 usage. Training an AI at the computing level of a human brain for one year can cost 126,000 L of water. Each year the computing power needed to train AI increases tenfold, requiring more resources. Water use of big tech companies' data centres is grossly underestimated—for example, the water consumption at Microsoft's Dutch data centre was four times their initial plans. Demand for water for cooling will only increase because of rising average temperatures due to climate change.

A similar case exists for all other aspects of resources that humans heavily rely on, for which there is plenty of literature available. A detailed scrutiny of them would bring critical sustainability issues to light but unfortunately, with the exponential growth [36] in the size of models, as well as the associated increase in computational complexity and rapid growth in the volume of health data, the development of accurate AI/ML models, which often consume extensive resources for training as well as testing, is only doubling down on its consumptions of resources in relation to storage, computing power, networking and domain expertise.

4 Self Driving Cars

Self-driving vehicles provide the transportation capabilities of conventional vehicles but are largely capable of perceiving the environment and self-navigating with minimal or no human intervention [37]. The beginning of the Second World War saw the concept of 'self-driving car' emerge for the first time from USA and the era of the mid-2010s, gain traction in the autonomous car industry through huge investments flowing in from big-time players such as Ford and BMW. The twenty-first century has been a witness to the rise of corporate giants—such as Tesla—coupled with a proliferation in the need for reduced reliance on hydrocarbon, which has resulted in an unprecedented rise in popularity of electric vehicles not just in the (developed) West but also in the (developing) global South. With international organizations showing their support for universally adopting the above-mentioned initiative, the question is that of "when", instead of "if and how", as regards its adoption. Though these innovative systems appear to be more efficient than conventional vehicles, the powering of their accessories—from cameras and radars to different sensors and the engine itself—have huge computational needs. As per an estimate, a production car with just cameras and radar generates close to 6 gigabytes of data every 30 s, which grows higher for a self-driver having additional sensors such as LIDAR. Modern vehicles are presently dependent on radar sensor inputs for inter-vehicle distance data and driver assistance systems [38]. This includes combining, sorting, and converting data into a robot-friendly picture of the world, with instructions on navigation needs, all of which requires vast reserves of computing power [39].

Here are some terminologies that we must know.

4.1 Connected and Autonomous Vehicle

A Connected Autonomous Vehicle (CAV) is that which can replace the driver for some or all the driving tasks [40]. This is a new, rapidly emerging domain, which has created a complete industry within a mere decade [41]. This includes vehicles acting automatically on brakes, the accelerator and/or the steering control.

4.2 Fully Automated Vehicle

They are vehicles that operate on automation, which is function-specific. In other words, it is centred on handling issues such as speed control, but leaves broader decision-making to the driver [42]. Automation advancements in the vehicle industry is therefore a continuous process that delivers several benefits such as: reduced driver fatigue, improved road safety, greater fuel efficiency, smart parking options etc. [43]. To be able to barely understand all CAV energy impacts, a peek into human behaviours and infrastructure systems merits a demand. Of the many categories of impacts, four of them are [44]—(1) Travel Demand Change; (2) Driving Details; (3) Multi-Vehicle Operation; (4) Power train Energy Source.

The existing literature on connected and autonomous vehicles mostly addresses their potential impact on the likelihood of traffic safety, travel behaviour and congestion, as well as energy use. The effects of partially to fully automated vehicles on traffic performance and greenhouse gas emissions are still obscure [45]. Nonetheless, an earnest attempt to quantify environmental impacts of the consumption of this resource has been made through credible references in the text. Shown below is an objective overview of each of these categories, along with their intrinsic features and expected outcomes.

4.2.1 Travel Demand Change (a Possible Changes in Users' Travel Behaviour)

1. Opting for a Better Route

Its imminent qualities would include an immediate and better judgment of traffic conditions in real time; identification of path(s) with fewer stops; and facilitation of dynamic ride-sharing (DRS) with strangers casually carpooling together en-route. As a result, studies done by Wadud et al. [46] state that the energy consumption would see a reduction in enhanced route choices from as little as -5% to as much as -20%. Wadud et al. 46 mention eco-driving, traffic smoothing, and vehicle light weighting as potential mechanisms by which automation could reduce energy consumption and emissions per kilo meter.

2. Newly Induced Trips from Under-served Population

It is the category of senior adults and persons with disabilities that mainly comprises of under-served populations in any geography. That, asides doing away with driver's licenses while riding in CAVs; and improved accessibility, constitutes the primary features of the aspect in focus. As a result, studies done by Wadud et al. [46] project an increase in energy consumption from newly induced trips, to be somewhere between 10 and 14%. Wadud et al. [46] note the potential for reduced per-km emissions and increased travel demand with automated vehicles, but offer few numbers. Their discussion of travel demand effects focuses mainly on the effects of extending mobility to underserved groups and induced demand from capacity improvements.

3. Increased VMT (vehicle-miles travelled) and Empty Driving

Owing to the rapid proliferation of shared automated vehicles (SAVs) which must also account for empty driving in between passengers, an estimate by [46] sees the increase in energy consumption from SAVs without dynamic ride sharing—to be between 6 and 14%—compared to the current state of vehicular energy consumption.

4. Long-distance Travel with CAVs

As per Wadud et al. [46] intra-urban and inter-regional travel would increase by 9.6% with the adoption of personal CAVs that would even affect competing travel modes such as airline travel. Therefore the increased travel distance with CAVs would increase energy consumption from 6 to 18% for ground transportation, as per [46].

4.2.2 Driving Details (the Possible Changes in Vehicle Performance)

Improved Driving Skill (leading to faster travel):

This would ensure driving at faster speeds, while ensuring safety that may also lead to an increase in speed-limit on certain roadways. Future road infrastructures would therefore, have a larger capacity compared to that of the current paradigm. In light of what is stated above, Wadud et al. [46] estimate a change in energy consumptions from 7% to as much as 30% for faster travel speeds of CAVs, as compared to what currently prevails.

1. Smoother Driving Cycle

A very logical expectation in this regard is the removal of redundant driving cycles that increase fuel consumption and are unfavourable for vehicle durability. It has therefore been determined by Wadud et al. [46] that fully automated vehicles have lower emissions than noisy driving cycles that are commonly found in non-automated driving methods. The reason being, as per Barth et al. (2009), reduced fuel consumption in dynamic eco-driving strategy of human drivers that reduces fuel consumption and lowers CO2 emissions. Future trends as viewed by Liu et al. (2017), Barth et al. (2009), that put energy consumptions (from smoother driving cycles) togo down to −10 and −20% of current energy use, are signs of optimism.

2. Computer and Sensor Power Demands

Because such an arrangement is expected to deliver a better performance and an improved driving experience, the energy use is bound to vary as per the level of automation. Going by the assessments of , Wadud et al. (2016), a CAV system is determined to have, at most, 4% greater emissions during its lifetime compared to a vehicle without this system in a life-cycle assessment analysis, albeit the actual energy consumption on the road could exceed this estimate [46]. This could get a CAV system to increase the energy consumption from 4% to as much as 15% [46].

Assumptions [for the analysis]: (a) 1000 watts of auxiliary load (for steady-state air-conditioning); (b) 1000 Watts of normal usage (high performance CPU and GPU equivalent).

4.2.3 Multi-Vehicle Operation Category (Includes Interaction Among Multiple CAVs and Infrastructures)

1. Vehicle-to-vehicle (V2V) and Platooning

This essentially means communicating with other vehicles through networking technologies which would certainly rely on better velocity synchronization and closer spacing among vehicles that could lead to a lesser aerodynamic drag on each unit, of which, a direct outcome is an experience of driving with reduced acceleration noise.

2. Vehicle-to-infrastructure (V2I) and Connected Intersection

V2I is like V2V, yet different because the communication here is with, for instance, intersections; lane markers and parking meters, viz., infrastructure. This would assist in reducing vehicle-stop delay , which places the expected energy reduction from V2I between 13 and 44%, as stated in studies conducted by Wadud et al. [46].

4.2.4 Power-Train Energy Source Category (Outcome of a New Power Source that Fuels the Transition to Electric Vehicles)

1. Electric and Hybrid Electric Vehicles

Forecasting long-term CAV technologies' adoption is not easy: many demand-side factors (e.g., willingness to pay [WTP]) and supply-side factors (e.g., technology prices) must be considered. Predictions on upcoming adoption rates are mostly based on the extrapolation of trends associated with previous vehicle technologies, expert opinions, or forecasts of supply-side variables. But a study [47] incorporates a formal mechanism to anticipate Level 1 and Level 2 automation adoption rates (including lane-centering assistance and adaptive cruise control, for example) and/or vehicle connectivity (using dedicated short-range communications [DSRC] technologies) in United States of America. The simplest of all awareness is that more vehicles equipped with electric motors would be a common sight to behold in the future.

The year 2015 saw almost 1.3 billion vehicles in use worldwide, of which nearly 77% were passenger vehicles (OICA 2021), a figure that may double by the end of the 2020s or early 2030s based on current trends, as per [46]. This makes the private cars—passenger cars/SUVs/small vans—account for approximately 60% of greenhouse gas (GHG) emissions [48]. GHGs produced by fuel sources such as diesel, natural gas, and gasoline are one of the leading causes of the greenhouse effect worldwide [49] simply due to: (1) rich emissions of methane (CH_4), carbon dioxide (CO_2) and nitrous oxide (N_2O) which have an impact potential that spans for decades [50]; (2) their ability to disturb natural gas cycles governing the planet and posing as significant threats to various flora and fauna types [51]. Thus, the seemingly virtuous act of introducing EVs is fraught with its own share of vices, particularly due to the resources that they consume for infrastructure (manufacturing + maintenance), on-board equipment, edge computing abilities (through edge servers), vehicle servicing, etc. Some studies have also argued in favour of the neutralisation of emission reduction benefits by AVs, considering increased travel due to lower costs involved, thus encouraging additional travel, and leading to an energy "rebound effect" [45]. This might risk the possibility of stepping into unsustainable waters sooner than the estimates discussed at intergovernmental meetings.

The newly released EU taxonomy is an ambitious and significant regulation aimed at defining sustainable activities. An activity qualifying as sustainable and in alignment with the EU taxonomy must not cause significant harm (DNSH) to the environmental objectives, namely (i) climate change mitigation, (ii) climate change adaptation, (iii) sustainable use of water resources, (iv) transition to a circular economy, (v) pollution prevention, and (vi) protection and restoration of biodiversity and ecosystems [52]. The DNSH principle was established as a qualifying element for the eligibility of an investment for the purpose of accessing better financial instruments in terms of conditionality and costs [52]. While understanding the impact of AVs on each of the components of the DNSH is crucial to ensure responsible and sustainable implementation, a holistic understanding of AVs and the development of an integrative vision incorporating pro-active mitigation measures is a topic that is yet to show up on the international research agenda [53].

5 Added Pressure on REEs

Rare earth elements (REEs) are important for green and a large variety of high-tech technologies [54] and are, therefore, in high demand. Due to the multi-dimensional nature of their applications, coupled with metrics of low supply and high demand, (REEs) are rightfully conceived as the most critical chemical elements [55] present on earth. As a result, supply with REEs is likely to be disrupted (the degree of which depends on the REE) in the near future [56]. They comprise of the lanthanide series elements, aka., the f-block elements of the modern periodic table and have currently become very critical to several modern technologies ranging from cell phones and televisions, to even LED light bulbs and wind turbines. Some examples on how these REEs are used in our daily life are—(1) Medical Science: Portable X-ray machines, X-ray tubes, magnetic resonance imagery (MRI) contrast agents, nuclear medicine imaging, cancer treatment applications, and for genetic screening tests, medical and dental lasers; (2) Renewable Energy: Hybrid automobiles, wind turbines, next generation rechargeable batteries, bio fuel catalysts; (3) Manufacturing: Hybrid automobiles, wind turbines, next generation rechargeable batteries, bio fuel catalysts; (4) Electronics: Television screens, computers, cell phones, silicon chips, monitor displays, long-life rechargeable batteries, camera lenses, light emitting diodes (LEDs), compact fluorescent lamps (CFLs), baggage scanners, marine propulsion systems, etc.

The estimated average concentration of the REE in the Earth's crust which ranges from around 130–240 μg/g [57], is significantly higher than other commonly exploited elements, and much higher than their respective chondritic abundance. The security of their supply has been a concern for import-dependent industrialized countries that have ambitions of advancing their low-carbon economies [57] (Table 1). Other aspects such as their applications in agriculture, medicine together with the environmental effects and search for alternatives to REE in different applications are also presented.

The 17 REEs are divided into heavy and light REEs [56]. China is the leading country in the production of REEs, having provided since recently more than approximately 90% of the global REE supply as per Long (2011). Facts stated in studies done by Wang and Liang (2014), talk about human exposures to REEs occurring due to anthropogenic activities in different ways such as large scale mining, mismanagement of e-waste, REE-enrichment of atmospheric particulates by strong winds have fallouts which can lead to some seriously disturbing trends among diverse geographies.

A major drawback with excessive REE extraction is that of the set of devastating health effects faced by on-field and off-field workers. Even for above average exposures through food chains, lifestyle and occupational choices, inhalation, ingestion, skin contact, pesticides, herbicides, fertilizers, soils and even their roots, the following concerns have started showing signs of commonality—(1) high urine-REE and blood-REE concentrations of people living near mining sites [59]; (2) non-cancerous toxicity [59]; (3) the possibility of REEs playing a role in the development of anaemia, as per La. Henríquez-Hernández et al. (2017), [59]; (4) excessive mortality due to lung and bladder cancer, as per De Vathaire et al. (1998), [59]; (5) accumulation of REEs in the cerebral cortex by long-term environmental and cause subclinical damage, as per Zhu et al. (1997), [59]; (6) lower levels of total protein, globulin, albumin, and serum-glutamic pyruvic transaminase (SGPT) and higher levels of IgM in blood serum, as per Zhu et al. (2005) [59]; (7) brain-tumor tissues of patients with astrocytoma, as per Zhuang et al. (1996), [59]; (8) illnesses such as indigestion, diarrhoea, abdominal distension, anorexia, weakness, and fatigue, coupled with lower levels of total protein, globulin, albumin, and serum-glutamic pyruvic trans-aminase (SGPT) and higher levels of IgM in blood serum, as per Zhu et al. (2005), [59]; (9) low levels of calcium in hair and a high risk of hypertension among housewives, as per Wang et al. (2017), [59]; (10) association of women undergoing IVF embryo transfers with a 30% increase in clinical pregnancy failure, a 230% increase in preclinical spontaneous abortion, and a negative correlation with the number of good-quality oocytes, as per Li et al. (2021), [59]; (11) high concentrations of La, Ce, Eu, and Gd in the semen from drinkers and long-term smokers, as per Marzec-Wroblewska et al. (2015), [59] etc.

So far the feasibility and potential yield of REE recycling has not been systematically evaluated [60] as the world has predominantly been focusing on reducing supply dependencies from China, the dominant supplier, through baby steps on end-of-life recycling, as promoted by OECD countries in the current global supply crisis.

6 Conclusion

There are two major realities that the world at large must come to terms with. The first, being that of the nearly inseparable co-existence between artificial intelligence systems (along with their supporting modules) and modern-day humans, with greater skewness of mutual dependencies on AI rather than that on humans. Regardless of our agreement on current choices of lifestyle or work style, the stakes of a 'computer with a bunch of fairly intelligent codes' are higher than what can be comprehended. Many professional and personal decisions now rely on machine-based decisions over human judgement, making it difficult to abandon its utility without a futuristic solution. The other reality is the message that this chapter has been trying to convey to its readers, who by now, have been able to comprehend the spirit of the text. In simple words, it encourages people to adopt the most responsible course of action, which is to step out of their newly found comforts residing within a machine's abilities.

As regards their scales of damage, weapons of mass-destruction and AI systems are fairly comparable across temporal scales of a few seconds, and a few decades, respectively. But using AI as an instrument to fire nuclear weapons at will, would be the most undesirable outcome that lies within the scope of our imagination. This is, but only one dimension i.e., peaceful coexistence, that forms a part of Musk's inevitable truth. Dimensions of environmental degradation, plummeting socio-economic development (especially in rising economies such as India and Africa) and hampered human development (starting with over-dependence of children on AI for meagre purposes) too are pernicious influences thriving even in the third decade of the twenty-first century. Their consequences are equally catastrophic, and should not be lost track of.

The way forward lies in striking a careful balance between the paradoxes of using AI for 'unavoidable and beneficial purposes'. World-leaders in international summits on climate change, adaptation (of environment-friendly approaches to absorb the present damage) and mitigation (of non-negotiable vices) have encouraged nations to aim for a 'net-zero emitter' status in their use of hydrocarbons, instead of rebuking their dependence on fossil fuels. Taking a cue from this philosophy, it would be prudent upon the modern world to meet its developmental aspirations with the help of AI, while simultaneously taking voluntary measures to balance out all detrimental aspects of the same. A place to start would be at the domestic level through induced checks and balances; supervised usage and a minimum threshold age-of-use for minors; imposing restrictions wherever possible; and so forth. Governments and educational institutions would have an important role to play in the enforcement of such commensurate policies and the dissemination of awareness, respectively. Certainly, a scientific methodology of measuring computed development against the efficacy of our internal policies would become an inevitable necessity for knowing the financial and human capital to be invested, for having a net zero effect on our environment, within the margin of error. As the fight against environmental degradation can have no by-standers, therefore, the world in general and the global scientific community in particular, shouldn't wait long for all the other detrimental factors to unfold, before we are left with no choice but to rely on what fate would have in store for us.

References

1. Arthur, W. B. (2010). *The nature of technology: What it is and how it evolves*. Penguin UK.
2. Galaz, V. (Ed.). (2014). *Global environmental governance, technology and politics: The Anthropocene gap*. Edward Elgar Publishing.
3. Westley, F., Olsson, P., Folke, C., Homer-Dixon, T., Vredenburg, H., Loorbach, D., & Van Der Leeuw, S. (2011). Tipping toward sustainability: Emerging pathways of transformation. *Ambio, 40*, 762–780.
4. Zhong, S., Zhang, K., Bagheri, M., Burken, J. G., Gu, A., Li, B., Zhang, H., et al. (2021). Machine learning: New ideas and tools in environmental science and engineering. *Environmental Science & Technology, 55*(19), 12741–12754.
5. Bhatnagar, R. (2018). Machine learning and big data processing: a technological perspective and review. In *The International Conference on Advanced Machine Learning Technologies and Applications (AMLTA2018)* (pp. 468–478). Springer International Publishing.
6. Surya, L. (2016). An exploratory study of machine learning and it's future in the United States. *International Journal of Creative Research Thoughts (IJCRT)*, ISSN 2320-2882.
7. Ferguson, N. S. (2014). Optimization: A paradigm change in nutrition and economic solutions. *Advances in Pork Production, 25*(1), 121–127.
8. Pomar, C., & Remus, A. (2019). Precision pig feeding: A breakthrough toward sustainability. *Animal Frontiers, 9*(2), 52–59.
9. Neethirajan, S. (2020). The role of sensors, big data and machine learning in modern animal farming. *Sensing and Bio-Sensing Research, 29*, 100367.
10. Javaid, M., Haleem, A., Singh, R. P., Suman, R., & Rab, S. (2022). Significance of machine learning in healthcare: Features, pillars and applications. *International Journal of Intelligent Networks, 3*, 58–73.
11. Raza, M., Sakila, K. S., Sreekala, K., & Mohamad, A. *Carbon footprint reduction in cloud computing: Best practices and emerging trends.*
12. Arora, A., & Basu, N. (2023). Machine learning in modern healthcare. *International Journal of Advanced Medical Sciences and Technology, 3*(4), 12–18.
13. Galaz, V., Centeno, M. A., Callahan, P. W., Causevic, A., Patterson, T., Brass, I., Levy, K., et al. (2021). Artificial intelligence, systemic risks, and sustainability. *Technology in Society, 67*, 101741.
14. https://www.theverge.com/24066646/ai-electricity-energy-watts-generative-consumption
15. Sasha Luccioni, A., Jernite, Y., & Strubell, E. (2023). Power hungry processing: Watts driving the cost of AI deployment? *arXiv e-prints*, arXiv-2311.
16. https://www.thirdway.org/memo/how-ai-uses-energy#:~:text=From%20Luccioni%20et%20al%2C%201%2C000,0.0047%20hours%20or%2017%20seconds.
17. Heikkilä, M. (2023). Making an image with generative AI uses as much energy as charging your phone. *MIT Technology Review, 1*, 2023.
18. https://www.datacenterdynamics.com/en/news/global-data-center-electricity-use-to-double-by-2026-report/
19. ShreyasMadhav, A. V., Rajaraman, R., Harini, S., & Kiliroor, C. C. (2022). Application of artificial intelligence to enhance collection of E-waste: A potential solution for household WEEE collection and segregation in India. *Waste Management & Research, 40*(7), 1047–1053.

20. Asante, K. A., Amoyaw-Osei, Y., & Agusa, T. (2019). E-waste recycling in Africa: Risks and opportunities. *Current Opinion in Green and Sustainable Chemistry, 18*, 109–117.
21. Heacock, M., Kelly, C. B., Asante, K. A., Birnbaum, L. S., Bergman, Å. L., Bruné, M. N., Suk, W. A., et al. (2016). E-waste and harm to vulnerable populations: A growing global problem. *Environmental Health Perspectives, 124*(5), 550–555.
22. Widmer, R., Oswald-Krapf, H., Sinha-Khetriwal, D., Schnellmann, M., & Böni, H. (2005). Global perspectives on e-waste. *Environmental Impact Assessment Review, 25*(5), 436–458.
23. Ruiz, A. (2022). *Latest global E-waste statistics and what they tell us*. TheRoundup.
24. Fawole, A. A., Orikpete, O. F., Ehiobu, N. N., & Ewim, D. R. E. (2023). Climate change implications of electronic waste: Strategies for sustainable management. *Bulletin of the National Research Centre, 47*, 147.
25. Oteng-Ababio, M. (2010). E-waste: An emerging challenge to solid waste management in Ghana. *International Development Planning Review, 32*, 191–206.
26. Debnath, B., Chattopadhyay, A. K., & Kumar, T. K. (2024). An economic optimization model of an e-Waste supply chain network: Machine learned kinetic modelling for sustainable production. *Sustainability, 16*(15), 6491.
27. The true cost of AI innovation: https://www.scientificcomputing.com/analysis-opinion/true-cost-ai-innovation
28. Kanungo, A. (2023). The green dilemma: Can AI fulfil its potential without harming the environment. *Earth.*
29. Bouza, L., Bugeau, A., & Lannelongue, L. (2023). How to estimate carbon footprint when training deep learning models? A guide and review. *Environmental Research Communications, 5*(11), 115014.
30. Vipra, J., Rodrigues, C., Bioni, B., Zanatta, R., Gillwald, A., Wu, B., & Lall, T. *Governing computational infrastructure for strong and just AI economies.*
31. Robbins, S., & van Wynsberghe, A. (2022). Our new artificial intelligence infrastructure: Becoming locked into an unsustainable future. *Sustainability, 14*(8), 4829.
32. Li, P., Yang, J., Islam, M. A., & Ren, S. (2023). Making AI less "thirsty": Uncovering and addressing the secret water footprint of ai models. arXiv preprint arXiv:2304.03271.
33. Gupta, J., Bosch, H., & Vliet, L. V. (2024). AI's excessive water consumption threatens to drown out its environmental contributions. *The Conversation.*
34. Erickson, P., Kartha, S., Lazarus, M., Tempest, K. (2015). Assessing carbon lock-in. *Environmental Research Letters, 10*, 084023.
35. OECD. (2014). *Energy, climate change and environment: 2014 Insights*. International Energy Agency.
36. Jia, Z., Chen, J., Xu, X., Kheir, J., Hu, J., Xiao, H., Shi, Y., et al. (2023). The importance of resource awareness in artificial intelligence for healthcare. *Nature Machine Intelligence, 5*(7), 687–698.
37. Yeong, D. J., Velasco-Hernandez, G., Barry, J., & Walsh, J. (2021). Sensor and sensor fusion technology in autonomous vehicles: A review. *Sensors, 21*(6), 2140.
38. Log, M. M., Thoresen, T., Eitrheim, M. H., Levin, T., & Tørset, T. (2023). Using low-cost radar sensors and action cameras to measure inter-vehicle distances in real-world truck platooning. *Applied System Innovation, 6*(3), 55.
39. Stewart, J. (2018). Self-driving cars use crazy amounts of power, and it's becoming a problem. *Wired Transportation.*
40. https://www.ferrovial.com/en/innovation/technologies/connected-autonomousvehicles/#:~:text=Connected%20Autonomous%20Vehicles%20(CAV)%20are,available%20on%20the%20EU%20market.
41. Takács, Á., Drexler, D. A., Galambos, P., Rudas, I. J., &Haidegger, T. (2018, June). Assessment and standardization of autonomous vehicles. In *2018 IEEE 22nd International Conference on Intelligent Engineering Systems (INES)* (pp. 000185–000192). IEEE.
42. Antsaklis, Panos J., Passino, Kevin, M.. Wang, S.J. (1991). *An introduction to autonomous control systems.*

43. Ahmed, H. U., Huang, Y., Lu, P., & Bridgelall, R. (2022). Technology developments and impacts of connected and autonomous vehicles: An overview. *Smart Cities, 5*(1), 382–404.
44. Lee, J., &Kockelman, K. M. (2019, January). Energy implications of self-driving vehicles. In *Proceedings of the 98th Annual Meeting of the Transportation Research Board, Washington, DC, USA* (pp. 13–17).
45. Massar, M., Reza, I., Rahman, S. M., Abdullah, S. M. H., Jamal, A., & Al-Ismail, F. S. (2021). Impacts of autonomous vehicles on greenhouse gas emissions—Positive or negative? *International Journal of Environmental Research and Public Health, 18*(11), 5567.
46. Wadud, Z., MacKenzie, D., & Leiby, P. (2016). Help or hindrance? The travel, energy and carbon impacts of highly automated vehicles. *Transportation Research Part A: Policy and Practice, 86*, 1–18.
47. Bansal, P., & Kockelman, K. M. (2017). Forecasting Americans' long-term adoption of connected and autonomous vehicle technologies. *Transportation Research Part A: Policy and Practice, 95*, 49–63.
48. Silva, Ó., Cordera, R., González-González, E., & Nogués, S. (2022). Environmental impacts of autonomous vehicles: A review of the scientific literature. *Science of The Total Environment, 830*, 154615.
49. Facts, E. F. (2015). *US transportation sector greenhouse gas emissions 1990–2013.* US Environmental Protection Agency.
50. National Academies of Sciences, Engineering, and Medicine. (2016). *Commercial motor vehicle driver fatigue, long-term health, and highway safety: Research needs.* National Academies Press. ISBN 0-309-39252-7.
51. Metcalf, G. E. (2009). Market-based policy options to control US greenhouse gas emissions. *Journal of Economic Perspective, 23*, 5–27.
52. European Union. *Regulation EU 2020/852 of the European parliament and of the council establishing a framework to facilitate sustainable investment.* Available online: https://eur-lex.europa.eu/legal-content/IT/TXT/?uri=CELEX:32020R0852. Accessed January 28, 2023)
53. Luca, O., Andrei, L., Iacoboaea, C., & Gaman, F. (2023). Unveiling the hidden effects of automated vehicles on "do no significant harm" components. *Sustainability, 15*(14), 11265.
54. Dutta, T., Kim, K. H., Uchimiya, M., Kwon, E. E., Jeon, B. H., Deep, A., & Yun, S. T. (2016). Global demand for rare earth resources and strategies for green mining. *Environmental Research, 150*, 182–190.
55. Tunsu, C., Petranikova, M., Gergorić, M., Ekberg, C., & Retegan, T. (2015). Reclaiming rare earth elements from end-of-life products: A review of the perspectives for urban mining using hydrometallurgical unit operations. *Hydrometallurgy, 156*, 239–258.
56. De Boer, M. A., & Lammertsma, K. (2013). Scarcity of rare earth elements. *Chemsuschem, 6*(11), 2045–2055.
57. Pell, R. S., Wall, F., Yan, X., & Bailey, G. (2019). Applying and advancing the economic resource scarcity potential (ESP) method for rare earth elements. *Resources Policy, 62*, 472–481.
58. Balaram, V. (2019). Rare earth elements: A review of applications, occurrence, exploration, analysis, recycling, and environmental impact. *Geoscience Frontiers, 10*(4), 1285–1303.
59. Brouziotis, A. A., Giarra, A., Libralato, G., Pagano, G., Guida, M., & Trifuoggi, M. (2022). Toxicity of rare earth elements: An overview on human health impact. *Frontiers in Environmental Science, 10*, 948041.
60. Rademaker, J. H., Kleijn, R., & Yang, Y. (2013). Recycling as a strategy against rare earth element criticality: A systemic evaluation of the potential yield of NdFeB magnet recycling. *Environmental Science & Technology, 47*(18), 10129–10136.

MIX
Papier aus verantwortungsvollen Quellen
Paper from responsible sources
FSC® C105338

If you have any concerns about our products,
you can contact us on
ProductSafety@springernature.com

In case Publisher is established outside the EU,
the EU authorized representative is:
Springer Nature Customer Service Center GmbH
Europaplatz 3, 69115 Heidelberg, Germany

Printed by Libri Plureos GmbH
in Hamburg, Germany